The Development of the Mechanics' Institute Movement in Britain and Beyond

The Development of the Mechanics' Institute Movement in Britain and Beyond questions the prevailing view that mechanics' institutes made little contribution to adult working-class education from their foundation in the 1820s to 1890. The book traces the historical development of several mechanics' institutes across Britain and reveals that many institutes supported both male and female working-class membership before state intervention at the end of the nineteenth century resulted in the development of further education for all.

This book presents evidence to suggest that the movement remained active and continued to expand until the end of the nineteenth century. Drawing on historical accounts, Walker describes the developments which shaped the movement and emphasises the institutes' provision for scientific and technical education. He also considers the impact that the British movement had on the overseas development of mechanics' institutes – particularly in Canada, America, Australia and New Zealand. The book concludes with a discussion of the legacy of the movement and its contribution to twentieth-century adult education.

The Development of the Mechanics' Institute Movement advances the argument that the movement made a substantial contribution to adult education for the working classes and provided a firm foundation for further education in Britain and beyond. It will appeal to academics, researchers and postgraduate students in the areas of education, history and sociology, as well as the philosophy of education, technical and vocational education, and post-compulsory education.

Martyn Walker is a Principal Lecturer and researcher at the University of Huddersfield. He is a member of the Policy Research Group in the School of Education and Professional Development and is a member of the editorial board of the *Journal of Educational Administration and History*. Martyn is a member of the Royal Society of Arts and is a Senior Fellow of the Higher Education Academy. His work, based on the history of adult education, has been widely published in scholarly journals.

Routledge Research in Education

The Development of the Mechanics' Institute Movement in Britain and Beyond

Supporting further education for the adult working classes

Martyn Walker

Routledge
Taylor & Francis Group

LONDON AND NEW YORK

First published 2017
by Routledge
2 Park Square, Milton Park, Abingdon, Oxon OX14 4RN

and by Routledge
711 Third Avenue, New York, NY 10017

First issued in paperback 2017

Routledge is an imprint of the Taylor & Francis Group, an informa business

British Library Cataloguing in Publication Data
A catalogue record for this book is available from the British Library

Library of Congress Cataloging in Publication Data
Names: Walker, Martyn, author.
Title: The development of the mechanics' institute movement in Britain and beyond: "a practical education in reach of the humblest means" / by Martyn Walker.
Description: Abingdon, Oxon; New York, NY: Routledge, 2017.
Identifiers: LCCN 2016005149 | ISBN 9781138923553 (hardcover) | ISBN 9781315685021 (electronic)
Subjects: LCSH: Technical education–History–19th century. | Mechanics' insitutes–History–19th century. | Mechanics–Study and teaching–19th century. | Mechanics' institutes–Great Britain–History–19th century.
Classification: LCC T69 .W35 2017 | DDC 607.1/041–dc23
LC record available at http://lccn.loc.gov/2016005149

ISBN 13: 978-1-138-48957-8 (pbk)
ISBN 13: 978-1-138-92355-3 (hbk)

Typeset in Galliard
by Deanta Global Publishing Services, Chennai, India

Contents

Figures

Tables

Acknowledgements

I am indebted to many people in so many different ways in support of this book. In particular, Vicky Duckworth, who not only encouraged me to seriously consider writing up my research but also provided me with much-needed advice as to how to 'get started'. Without her 'nagging' the work would never have been written up. I would also like to thank Josephine Maltby, Professor of Finance and Accounts at Sheffield University, for her advice with regard to savings banks and penny banks. Garry Lucas, Professor of Hydraulics and Rob Brown, Professor of Chemistry, both of who work at the University of Huddersfield, for their expertise with technical issues associated with their respective disciplines. I would also like to thank staff at the Lancashire Public Records Office, Newcastle-upon-Tyne, Bradford, Leeds, Burnley, York, Liverpool, Manchester, Nottingham and Birmingham Libraries. I would like to thank Judith Taylor at NCFE, Newcastle, who arranged for me to use former Newcastle Institute annual reports and documents associated with the Northern Union of Mechanics' Institutes. I would also like to thank librarians at the University of Huddersfield particularly with regard to inter-loan resources, as well as advice received from Jim Lowden regarding the institutes of Australia, Adam Nelson, Professor of Education Policy and History at Wisconsin-Madison University, with regard to American institutes and Alistair Black, Professor of Library Information, at the University of Illinois in relation to the history of British libraries. The high quality maps produced from my data would have been impossible without the expertise and support of Keith Hartley, technician, in the School of Education and Professional Development at the University. I appreciate the support given by the University of Huddersfield, my employer, for research and scholarship time to complete the project, and by Roy Fisher in particular, who followed my progress with interest and gave useful advice. Last, but by no means least, I would like to thank my family for their patience and fortitude, particularly my wife, Hermione, during the writing of the book.

1   Introduction

The history of education

As way of a general introduction, it is important that the history of education as an academic subject is first discussed in order to put this study in context. As Crook and McCulloch (2002: 397) state, 'we need constantly to be reminded as educational historians, of the value of adopting a comparative approach'. As an academic discipline, the history of education provides insight into the educational past through comparisons and contrast with contemporary educational developments (ibid.: 2002). Any study relating to the history of education inevitably involves crossing several academic disciplines, most obviously history, education, politics and sociology. The strength of this has always been that eminent historians, educationalists and sociologists have all made substantial contributions to the subject of the history of education. Lowe (2002) states that the history of education should be central and not a peripheral subject and that educational historians have much to say on all sorts of issues relating to education. In many ways, the subject has often been found hidden behind the other disciplines.

Traditionally, the history of education has been included in textbooks on social and economic history with a strong coverage of school-age and university developments but with little reference to further or adult education. This is of much concern, as generations of children and young people are not being made aware of post–school education developments, or, if they are, only at a superficial level. Yet such studies relating to nineteenth-century and early twentieth-century adult education provide a wealth of knowledge and relevance to post–school education and training, as this book attempts to do within the context of the Mechanics' Institutes Movement.

The sociologist Durkheim put it succinctly when he stated in the early twentieth century that by 'studying the past it is possible to anticipate the future and understand the present'. In particular, he argued that,

> Only history can penetrate under the surface of the present educational system; only history can analyse it; only history can show us of what elements it is formed, on what conditions each of them depends, how they are interrelated; only history, in a word, can bring us to the long chain of causes and effects of which it is the result (McCulloch 2012: 26).

Briggs (1972: 14), writing from the historian's point of view, states that,

> History has always been a borrower from other disciplines and in that sense
> social-scientific history is just another example of a time-honoured process;
> but history had always been a lender, and all the social sciences would be
> immeasurably poorer without knowledge of the historical record. The social
> sciences are not a self-contained system, one of whose boundaries lies in
> some fringe area of the historical sciences. Rather the study of man is a con-
> tinuum, and social-scientific history is a bridge between the social sciences
> and the humanities.

Briggs believes that the study of the history of education should be considered
as part of the wider social history of society, involving the study of politics, economics, social class, feminism, culture and religion. He is particularly concerned
that the history of education could easily be seen as a sub-branch of history and
become increasingly separated. Sanderson (2007) highlights the work of Peter
Mathias, who, in 1971, gave an address at Oxford University, highlighting the
increasing close links between economic history and other disciplines or 'neighbours', as he referred to them as, such as law, sociology, demography and education. Simon (1960) believed that the main tasks of a historical study were to trace
the development of education and to try to assess the function it has fulfilled at
different periods, and 'to reach a deeper understanding of the function it fulfils
today' (McCulloch 2012: 27).

History of education studies

For many years the history of education was taught as a subject as part of teacher
training programmes which Richardson (1999) says had an emphasis on the
study of famous and influential educators and legislation of previous decades,
identifying their contributions and impact on the primary and secondary education sectors for those entering the teaching profession. The introduction of the
primary and secondary Bachelor of Education Degrees (BEd) in 1963 had the
four disciplines of history of education, philosophy, psychology and sociology as
core subjects underpinning teacher training. Aldrich (2003) noted that during
the 1970s and 1980s the curricula for secondary teacher training at the London
Institute of Education was based around these four core subjects, which were
distinctive elements of the qualification through supporting education theory and
practice as well as informing trainee teachers embarking on their future teaching
careers.

Government policy in Britain, however, in recent times has greatly influenced
(many would say interfered with) the training of teachers. The 'ology' academic
disciplines had become mistrusted for many years by political policymakers, particularly on the Right, who seemed to regard teacher training in universities as a
'hotbed' for radicalisation of the teaching profession. Unfortunately, with changes
in the qualifications framework, based more on reflective practice, pedagogical

underpinning and classroom-based observations, the history of education, along with sociology, psychology and philosophy of education, was also a casualty. The four-year BEd degrees, always treated with suspicion by Russell Group universities, were therefore replaced with the Postgraduate Certificate in Education when teaching became an all-graduate profession. This resulted in education topics and developments often overlooking historical perspectives, and the history of education no longer had the same status it once had.

In America, Franklin and Ortiz (2012) found that, despite its longevity in the curriculum, university historians have always been sceptical about its status and academic standing, particularly in relation to teacher training and education studies. Similarly in Britain, twentieth-century education departments were rather conservative, they lacked academic confidence and were viewed by history departments as being more involved in teacher training rather than academic research. This was due to educationalists previously having had experience as teachers in schools and lecturers in colleges, unlike historians who generally completed their Doctor of Philosophy Degrees (PhDs), before becoming postgraduate researchers and then taking up academic teaching posts in universities. This is reflected in the fact that while British historians were writing for academic journal publications from as early as the 1870s, the first education journal, the *British Journal of Educational Studies*, was only published in 1952 (Richardson 1999).[1]

However, educational historians continued to 'preach the word', albeit on a much smaller scale. The author, for example, delivers a lecture to the full-time one-year Postgraduate Certificate in Education (PGCE) course for students training to be teachers and lecturers for the further education sector, on the history of adult education. It is always well received. Aldrich (2003: 137) noted that at 'the first lecture of the year for PGCE secondary trainees at the Institute of Education in London was the History of the English Education System which regularly receives high ratings'. However, those wishing to qualify to teach in primary, secondary or further education are now spending more and more of their initial training period in schools and colleges, rather than at university, and as such 'are unlikely to acquire any substantive grounding in in the disciplines of sociology, psychology, philosophy and history' which until the 1990s were embedded in teacher training programmes in relation to education (Deem 1996: 148).

British university schools of education, with the decline in teacher training programmes, now offer education studies at first and higher degree with several delivering their own PhD and Doctor of Education Degree programmes (EdD). Thus, the history of education as a discipline has been thrown a lifeline, keeping it to the fore as a subject in relation to history, social science, social policy and education. Aldrich (2003: 134) rightly argues that all historians of education have a duty to both history and education and that this duty to the discipline of history is as great as that of the historian of politics, foreign affairs or family history. He further states the importance that 'the historian's prime duty is to record and interpret the events of the past for contemporaries and for future generations'. Aldrich states that historians often acknowledge the impossibility of discovering the absolute truth but this should not absolve them from searching for it

(McCulloch and Watts 2003). Laslett (1987: 264) makes reference to duties expected of historians when he stated,

> The first is a duty to his [sic] own generation and second to the people of the past. The third is shared not only with other historians and social scientists but with all scholars and scientists. This is the duty to search after the truth to the utmost of his capacity, or of hers, recognizing that it may be impossible to avoid some degree of bias but doing all that can be done to avoid it.

McCulloch (2012: 38) makes the crucial point that the subject of education has largely ignored historical aspects not only in Britain but also overseas. Indeed, both education studies and educational research offered in many higher education institutions rarely seems to include historical perspectives. As he puts it, 'in the spirit of Durkheim and Simon, it is urgent that historians of education consider new ways of promoting connections between historical insights and interpretations on the one hand and the needs of education on the other'. As Lowe (2002: 502) summarises, 'if we remain true to the canons of excellence and expertise ... we have much to say to educators, that we can inform and influence both policy making and practice' through the history of education. This book attempts to do just that through the developments and contributions that mechanics' institutes made to adult education in Britain and beyond, through continuity and changes over time.

Early developments

As early as the eighteenth century, with the emerging scientific discoveries and technological developments such as the steam engine, there was huge interest in popular science or, as it was referred to at the time, experimental philosophy. There was also a gradual development in more general education for the 'gentling masses'. Charity schools and Sunday schools were providing the three 'Rs' of reading, writing and arithmetic for children, being extended during the early nineteenth century through the monitorial system, and it would only be a matter of time before adult education had to be established more formally.

The catalyst for adult education was the Industrial Revolution with large numbers of the labouring population leaving the countryside, with its own culture and skills, and moving into the growing towns to work in the mills. Changes in working practices relating to a rigid routine not influenced by the seasons but formal hours and disciplined life in the mills, meant that middle-class educationalists thought more formal instruction was required based on the three 'Rs'. The Rev. Dr Richard W. Hamilton, President of the Leeds Mechanics' Institute and local Independent minister, observed in 1845 that the old 'pastimes of village buffoonery and rudeness were quite unfit for an age in which the marvels to technology and science were daily more apparent' (Hamilton 1845: 118).

During the first twenty years of the nineteenth century there was an emergence of several types of educational establishments, often referred to as mutual improvement societies, which were seen as important forms of what was termed working-class self-help and mental improvement. They flourished throughout Britain and particularly in Lancashire and Yorkshire. They varied in size but few had over 200 members by the mid-nineteenth century. The Mutual Improvement Society at Ramsbottom in East Lancashire, for example, had in 1850 about 180 members, of who fifty were attending classes (Watson 1989: 9). According to Radcliffe (1986), these societies brought working-class men and women together who attended classes of elementary subjects and later advanced discussions on a variety of subjects, for example those offered at Bradford Moor in the West Riding of Yorkshire which included such themes as metaphysics and philosophy (popular science). They seem to have been encouraged by religious groups, particularly the Methodists, referred to by Samuel Smiles, government reformer and author of *Self-Help* in 1859, as 'educational Methodism' (Harrison 1961: 50). Harrison (1961: 53) suggests that mutual improvement societies were 'the most truly indigenous of all early attempts at working class adult education'. They particularly concentrated in offering reading and writing in small towns and rural communities. Many pre-dated mechanics' institutes and were listed alongside other nineteenth-century educational institutions, often later changing their names and becoming mechanics' institutes (Hudson 1969: xi).

The catalyst for the founding of the Mechanics' Institute Movement was, first, the British Industrial Revolution, emerging from the late eighteenth century and which took hold from the 1800s onwards. It not only had a substantial economic impact on the whole country but also a social one. With technological developments there was a need, albeit small to start with, for adults having some understanding of rudimentary science, mathematics, English grammar and reading. Public lectures were usually based on scientific knowledge, out of which developed what became referred to as 'steam intellect' being offered by voluntary organisations, predominately the mechanics' institutes. Steam intellect became the link between work and education, a model of the steam engine sitting 'firmly on the lecturer's table' (O'Farrell 2004: 3).

Second was the growing interest in science, which was first delivered by itinerant lecturers travelling across the country to predominately middle-class audiences. In 1727, Glasgow University had opened its doors to the public, who could 'attend the lessons on experimental philosophy without a gown' (Kelly 1952: 18). This is reflected by the fact that science, referred to as 'useful knowledge', had less status than the classics undertaken by the privileged at public school and the ancient universities such as Oxbridge. Adam Smith, the eighteenth-century economist, stated in his book, *The Wealth of Nations,* first published in 1776, that geometry and science should be introduced to primary school age groups in support of Britain's ongoing economic success (Smith 1982 edition).

Hudson (1969: 29–30), writing in 1850 as a contemporary and supporter of mechanics' institutes, identified the various working-class institutions that had been established by the late eighteenth century including 'societies established for the dissemination of knowledge of the arts and sciences among the labouring population'. One of the members of a local branch of the Society, Thomas Clark, was giving lectures in his own house between 1794 and 1795, frequented by artisans, many of whom were employees of the Eagle Foundry in Birmingham.

They became known as 'the Cast Iron Philosophers' and were renowned as 'the best workmen' in the district.

In 1796 the Anderson Institution was founded by John Anderson, and in 1799 Dr George Birkbeck, a medical doctor and Quaker from Settle in Yorkshire, succeeded Dr Thomas Garnett, the first professor of natural philosophy at the Institution. Birkbeck gave lectures to working men, the first one being 'upon the mechanical properties of solid and fluid bodies', delivered in 1800. At about the same time, a group of working men in Birmingham were establishing a similar institution. Birkbeck lectured on chemistry, mechanics and electricity at various towns including those of Birmingham, Hull and Liverpool between 1804 and 1805 (Kelly 1952: 18).

Kelly (1952) suggests that the first Mechanics' Institute in Britain was the Edinburgh School of Arts which was opened in 1821, followed by a similar institute established in Glasgow having separated from the Anderson Institution in 1823, and took the name 'Mechanics' Institute', which was to become the most popular name for such foundations. In the same year, while Birkbeck is accredited with the establishment of the London Mechanics' Institution, there can be little doubt that other contemporaries supported the initiative, including his close friend, Henry Brougham. However, Birkbeck is seen as the founder of the Mechanics' Institute Movement.

A third factor that supported the establishment of the Mechanics' Institute Movement was that of social reform supported by, amongst others, the emerging evangelical Christian movements of the Unitarians and Quakers during the first half of the nineteenth century. Such philanthropists and humanitarians, who were mostly radical Whigs, supported abolition of slavery (1833), parliamentary reform (1832), temperance, factory reform, trade unionism and the Poor Law Amendment Act (1834). They also held strong beliefs in social justice and education for all. This was supported by Henry Brougham, who in 1825 wrote *Observations upon the education of the people* and in the same year established the Society for the Diffusion of Useful Knowledge. In 1837, he put forward a bill to Parliament supporting public education.

The fourth influence was 'history from below', that is, the labouring population aspiring to learn in support for both political reform and economic improvement. However, as O'Farrell (2004: 23) states, if the working classes 'were looking to the institutes to help it was an expectation that was largely unfulfilled'. The name 'mechanics' institute' was a self-inflicted hindrance from the start if it was attempting to attract the labouring classes to become members. Tylecote (1930) highlighted the confusion that contemporaries had with the term 'mechanic', which, with hindsight, she noted was probably not the ideal title. Indeed, historical debates have concentrated on the title, arguing that these institutes were not attended by mechanics and therefore were a failure in their objective of supporting working-class adult education.

Tylecote (1930: 265) gave the example of Ashton-under-Lyne, where Charles Hindley was President. He noted that the majority of members were 'young, between 18 and 25 years of age, but not strictly speaking, mechanics'. He believed that mechanics themselves misunderstood the function of mechanics' institutes, which was 'to advance their knowledge through scientific and philosophical education'. Hindley had interviewed some of them and their response to joining the Mechanics' Institute in Ashton was, '[w]hat? Do you think we want to have our trade taught to everybody at a rate of 2/6 per quarter? We are not such fools'.

Hindley made specific mention of the coalminers in the area when he wrote to Henry Brougham about his concerns that the title 'mechanic' was misleading and discouraged working men from attending.

> There is perhaps no class of English Society so degraded as the colliers and I felt very desirous that we should have some sort of their men as members. I called upon one and endeavoured to explain the advantages he would derive from attending the Institution – but no! The very term mechanic is enough for him and he insisted that he was no mechanic and could not be persuaded to interfere with the trade of another body of his fellow men (Tylecote 1930: 266).

Garner and Jenkins (1984: 144) highlight the difficulty of just which members of society made up the membership of institutes. They state that the word *mechanic* is a 'term with no fixed definition, often reserved for a man who worked with his hands, sometimes for a man who worked with machinery and often, somewhat pejorative, for a low workman'. They suggest the definition put forward by Harrison (1955) is probably an accurate one. For him, mechanics and artisans were skilled workers which distinguished them from the manufacturing or operative workforce in the mills who were semi-skilled or the labouring population who were unskilled.

Thus, the title 'mechanics' institute' was so misleading that it put off the group for which such institutions were designed, the working class. This indicates that the confusion of the title not only put off the labouring classes from attending, but also historians have been led to think that institutes failed because they did not attract the mechanics, a skilled upper working-class group. Either way, the specific title 'mechanics' institute' was a rather unfortunate one.

Mechanics' Institute Unions

Mechanics' Institute Unions were established which supported the sharing of teachers, library books and advice from visiting committee members from other institutes. Eight such unions were established across the country, the first one being the Yorkshire Union of Mechanics' Institutes which was founded in 1838, replacing the West Riding Union of Mechanics' Institutes which had been formed the year before. Katoh (1989: 6) states that Edward Baines organised and managed the Union and organised the publications of annual reports on all member institutes (or those that sent reports), and also arranged annual conferences, hosted by a different institute each year, at which knowledge and information could be exchanged through lectures and presentations. The union established guiding principles of management. Agents were employed to support all member institutes, and printed copies of the lectures were circulated so they could be read out in member institutes. The annual reports of the Yorkshire Union have been the basis for much of this research.

The Union of Lancashire and Cheshire Institutes was formed a year later, followed by the establishment over the next eight years of the Midlands, the Kent, the Northern, made up of Northumberland and Durham, and the Devon and Cornwall, with the Leicestershire Union being the last to be established in 1853 (Inkster 1976). All the Unions seem to have been organised on similar lines. The most successful were the Yorkshire Union, with the Leeds Institute being the headquarters, the Lancashire and Cheshire Union, based at the Manchester Institute, and the Northern Union with the Newcastle Institute as its headquarters. All three unions were well managed, supportive and adaptable to the needs of their membership. The Yorkshire Union of Mechanics' Institutes' annual reports are the most complete for any region in the British Isles and provide much insight into the Movement both regionally and nationally. Hudson (1969) and Kelly (1952) listed adult institutions such as mutual improvement societies, lyceums, scientific and literary institutions and working men's colleges alongside mechanics' institutes. They were all very similar in what they were offering, with mechanics' institutes being the most common (Hemming 1977).

Historical debates

Hudson (1969: xii–xiii), writing in 1851, stated that between 1849 and 1850 there had been an increase in the number of new or reopened institutes not seen since 1844. However, for 1851 (the year his book was published), there were no increases. He identified a number of factors for what he believed was the beginning of the end of the Movement despite fees being 'exceeding low'. First, he identified that the institutes should adapt to the changing needs of the time. This included the updating of news and reading rooms, libraries, classes and lectures, as traditional institutes that had not introduced newspapers and novels 'became extinct'. Second, Hudson noted that institutes had changed their priorities from supporting the labouring population, to 'unhealthy excitement by introducing weekly lectures, frequent concerts, ventriloquism and Shakespearian readings, directing their chief energies into a wrong channel (the middle classes) and involving the societies in debt and difficulty' (Hudson 1969: xii–xiii).

With regard to weekly lectures, Hudson had identified that the initial interest in science had been replaced with scientific theory that only the professional classes and well-educated people could understand and appreciate. Brougham (1825) also identified that mechanics and artisans were abandoning the mechanics' institutes. The findings by Hudson and Brougham have been taken as the key argument that the Mechanics' Institute Movement was a failure in its ambition to provide adult education to the masses. Kelly (1957) concludes his main body of research in 1851. Tylecote (1957: 291–2), writing in the same year as Kelly, but concentrating in more detail upon the textile districts of Lancashire and Yorkshire, concludes that: 'After twenty-five years of interesting and in many ways encouraging development the mechanics' institutes were, in 1850, looking forward hopefully to a further period of service under more promising social conditions with better educational foundations on which to build.'

While her work is very detailed and informative, second only to Kelly's research on the national movement, Tylecote concludes her research in 1850. In more recent years, Roderick and Stephens (1985: 123) argue that the mechanics' institutes were intended for the working classes but that the Movement was confined in the main to London, Scotland and the industrial north. They, like Kelly (1957) maintain that the Movement spread during the 1830s and 1840s, which they believed was the golden age, but that from the 1830s the mechanics institutes failed to meet the needs of the working class. 'Mechanics rarely attended ... teaching was unsystematic, activities had become frivolous and the libraries were only used for fiction.'

Shapin and Barnes (1977) state that there were 700 mechanics' institutes in England and Wales, with a total membership of 12,000, by 1851 and imply that this was the high point of their expansion. Wright (2001: 14) emphasises the points made by Hudson that 'they [institutes] had failed to attract the class whom they were intended to benefit'. Cotgrove (1958: 47) believed that mechanics' institutes failed and that 'they did not attract a majority of working men or teach much science'. This view has also been reinforced by historians such as Claeys (2000: 157) who has stated that

> mechanics' institutes were developed in the first half of the nineteenth century to further technical and adult education in Britain. Beginning in the early 1820s in Glasgow, Edinburgh, Leeds and London, there were about 700 mechanics' institutes and similar associations in Britain by 1850 ... they have often been accounted a failure.

Lucas (2004: 5), an educationalist, believes that the 'mechanics' institutes did not win credibility as genuinely mass adult education providers because their major emphasis was access to scientific knowledge through the reading of tracts and pamphlets and they assumed a high level of literacy'.

This volume attempts in the context of educational history to question previous research that suggests that mechanics' institutes were a failure by the 1850s and that they did not provide education for working-class adults, those for whom they were founded. Research will cover the period from circa 1821 to 1900, so as to include development, changes, external factors and policy which had an impact on the Movement until the end of the century, when technical schools and the beginning of further education was being developed for all.

Structure of this volume

This volume has been structured in such a way that it supports the key areas identified in Chapter 1 with regard to the subject of the history of education and the social sciences relating to key debates. Chapter 2 looks at the origins and history of the Mechanics' Institutes Movement and raises the crucial question that while historians have tended to concentrate on the period 1824 to 1850, what happened to mechanics' institutes after 1850? Chapter 3 explores the scientific education offered by the mechanics' institutes up to 1850, prior to the Great Exhibition of 1851 being opened in London, and the extent to which they were adaptable to the needs of their membership. Chapter 4 is concerned with scientific and technical education from 1850 until the end of the century, when technical schools and colleges were wholly state-funded. Chapter 5 provides an analytical overview of social class and membership relating to those who attended mechanics' institutes. Meanwhile, Chapter 6 discusses the contribution made by mechanics' institutes towards adult female education opportunities. Chapter 7 looks at the physical presence, through bricks and mortar, of mechanics' institutes from their origins to their later Victorian architecture. Chapter 8 is concerned with the opportunities mechanics' institutes gave their members with regard to providing access to newspapers, journals and books through their libraries. Chapter 9 covers extensive research into the Yorkshire Union with regard to the distribution of member institutes and detailed studies of three distinctive clusters. Chapter 10 is concerned with looking at the impact the British Movement had on overseas development of mechanics' institutes, particularly in Canada, America, Australia and New Zealand. Finally, Chapter 11 concludes with the legacy of the Mechanics' Institute Movement and its contribution to twentieth-century adult education.

Note

1 It is quite ironic that the *British Journal of Educational Studies* Volume 1, Issue 1, 1952 included an article by Thomas Kelly, who was at the time Director of Extra-Mural Department Studies at the University of Liverpool, entitled '*The Origins of the Mechanics' Institutes*'.

References

Aldrich, R. (2003) 'The three duties of the historian of education', *Journal of the History of Education Society* Vol. 33, No. 2, 133–143.

Briggs, A. (1972) 'The study of the history of education', *Journal of the History of Education Society* Vol. 1, No. 1, 5–22.

Brougham, H. (1825) *Practical Observations upon the Education of the People, Addressed to the Working Classes and their Employers* (London: Longman, Hurst, Orme, Brown and Green).

Claeys, G. (2000) 'Political economy and popular education: Thomas Hodgskin and the London Mechanics' Institute, 1823–1828', Davies, M. T. (ed.) *Radical and Revolution in Britain, 1775–1848* (London: Palgrave Macmillan): 157.

Cotgrove, S. F. (1958) *Technical Education and Social Change* (London: Allen and Unwin).

Crook, D., and McCulloch, G. (2002) 'Introduction: Comparative approaches to the history of education', *Journal of the History of Education Society* Vol. 31, No. 5, 397–400.

Deem, R. (June 1996), 'The future of educational research in the context of the social sciences: A special case?', *British Journal of Educational Studies* Vol. 44, No. 2, 143–158.

Franklin, B. M., and Ortiz, R. (2012) 'Educational history, policy research, and ethnohistory', Larsen, J. E. (ed.) *Knowledge, Politics and the History of Education: Studies in Education* Vol. 2 (Berlin: LIT-Verlag): 89–103.

Garner, A. D., and Jenkins, E. W. (1984) 'The English Mechanics' Institutes: The case for Leeds 1824–1842', *History of Education* Vol. 13, No. 2, 139–152.

Hamilton, R. W. (1845) *The Institutions of Popular Education* (Leeds).

Harrison, J. F. C. (1955) *Social and Religious Influences in Adult Education in Yorkshire between 1830 and 1870*, unpublished PhD thesis (Leeds: Leeds University).

Harrison, J. F. C. (1961) *Learning and Living, 1790–1960: A Study in the History of the English Adult Education Movement* (London: Routledge and Kegan Paul).

Hemming, J. P. (1977) 'The Mechanics' Institutes in the Lancashire and Yorkshire textile districts from 1850', *Journal of Educational Administration and History* Vol. 9, No. 1, 18–31.

Hudson, J. W. (1851) *The History of Adult Education in which is comprised a Full and Complete History of the Mechanics' and Literacy Institutions* (Reprint. London: Woburn 1969).

Inkster, I. (1976) 'The social context of an educational movement: A revisionist approach to the English Mechanics' Institutes, 1820–1850', *Oxford Review of Education* Vol. 2, 277–307.

Katoh, S. (1989) 'Mechanics' Institutes in Great Britain to the 1850s', *Journal of Educational Administration and History* Vol. XXI, No. 2, 1–7.

Kelly, T. (1952) 'The origins of Mechanics' Institutes', *British Journal of Educational Studies* Vol. 1, No. 1, 17–27.

Kelly, T. (1957) *George Birkbeck: Pioneer of Adult Education* (Liverpool: Liverpool University Press).

Laslett, P. (1987) 'The character of familial history, its limitations and the conditions for its proper pursuit', *Journal of Family History* Vol. 12, No. 1, 263–284.

Lowe, R. (2002) 'Do we still need history of education: is it central or peripheral?', *Journal of the History of Education Society* Vol. 36, No. 6, 491–504.

Lucas, N. (2004) *Teaching in Further Education: New Perspectives for a Changing Context* (London: Bedford Way Papers).

McCulloch, G., and Watts, R. (2003) 'Introduction: Theory, methodology, and the history of education', *Journal of the History of Education Society* Vol. 32, No. 2, 129–132.

McCulloch, G. (2012) 'The changing rationales of the history of education', Larsen, J. E. (ed.) *Knowledge, Politics and the History of Education: Studies in Education* Vol. 2 (Berlin: LIT-Verlag): 25–38.

O'Farrell, P. N. (2004) *Heriot-Watt University: An Illustrated History* (Edinburgh: Pearson Education Limited).

Radcliffe, C. J. (1986) 'Mutual improvement societies in the West Riding of Yorkshire, 1835–1900', *Journal of Educational Administration and History* Vol. 18, No. 2, 1–16.

Richardson, W. (1999) 'Historians and educationalists: The history of education as a field of study in post-war England 1945–1972, Part 1', *Journal of the History of Education Society* Vol. 28, No. 1, 1–30.

Roderick, G. W., and Stephens, M. D. (1985) 'Mechanics' Institutes and the State', *The Steam Intellect Societies, Essays on Culture, Education and Industry circa 1820–1914* (Nottingham: Department of Adult Education, University of Nottingham): 123.

Sanderson, M. (2007) 'Educational and economic history: The good neighbours', *Journal of the History of Education Society* Vol. 36, Nos. 4–5, 429–445.

Shapin, S., and Barnes, B. (1977) 'Science, nature and control: Interpreting Mechanics' Institutes', in *Social Studies of Science* Vol. 7, 31–74.

Simon, B. (1960) *Studies in the history of education 1780–1870* (London: Lawrence and Wishart).

Smith, A. (1982 edition) *The Wealth of Nations* Book V, Part III (Harmondsworth, England: Penguin English Library).

Tylecote, M. (1930) *The Mechanics' Institutes in Lancashire and Yorkshire, 1824–1850 with special reference to the Institutions at Manchester, Ashton-under-Lyne and Huddersfield*, unpublished PhD thesis (Manchester: University of Manchester).

Tylecote, M. (1957) *The Mechanics' Institutes of Lancashire and Yorkshire Before 1851* (Manchester: Manchester University Press).

Watson, M. I. (1989) 'Mutual improvement societies in nineteenth-century Lancashire', *Journal of Educational Administration and History* Vol. 21, No. 2, 8–17.

Wright, G. (2001) 'Discussions of the characteristics of Mechanics' Institutes in the second half of the nineteenth century: The Bradford example', *Journal of Educational Administration and History* Vol. 33, No. 1, 1–16.

2 A brief history of the Mechanics' Institute Movement

Introduction

Industrialists and manufacturers from at least the last decade of the eighteenth century were realising the importance of education and training with regard to both family and workforce, pre-dating the Mechanics' Institute Movement. Several organisations and societies were being established for this purpose. Dick (2008: 568) highlights the importance of one of these, the Lunar Society, which was 'a network of male scientists, industrialists and writers' who were all part of this 'energetic British philosophical society which symbolised enlightenment in science' in the Midlands between 1765 and 1791. Members included Matthew Boulton, who in 1762 opened his Soho Works, which manufactured a variety of products from buttons to steam engines, and James Keir, inventor and industrialist of glass manufacture, chemicals and mining. Other members included Thomas Day, radical writer and educationalist, Richard Edgeworth, inventor and educationalist, Joseph Priestley, experimental scientist, educationalist, political and religious campaigner, Josiah Wedgwood, the ceramics manufacturer, and James Watt, partner of Matthew Boulton and inventor who improved the efficiency of the steam engine (ibid.). Other scientists and manufacturers were in correspondence with members of the Lunar Society, including Benjamin Franklin and Joseph Wright, who regularly discussed and debated ideas with them.

Not only did such men share and pass on their knowledge and understanding amongst each other but also to their next of kin, such as Boulton working closely with his son, also Matthew, training him in geometry, drawing, mathematics, modern languages and scientific experiments, all important subjects for industry. Boulton also offered training to his workforce. Samuel Smiles, writing his book entitled *Lives of Boulton and Watt* in 1865, observed that 'Soho was spoken of with pride, as one of the best schools of skilled industry in England' (Smiles 1865: 23). Erasmus Darwin, himself closely connected with the Lunar Society, believed it to be very important that middle-class boys and girls had an understanding of industry as part of their education and should visit 'manufactories which adorn and enrich the country' during their school holidays (Dick 2008: 569). Thus, the need for an adult educated workforce had become apparent by the turn of the century and was much different from traditional subjects offered at school and university – the 'classics'.

As mentioned in Chapter 1 of this volume, the first mechanics' institute was founded in Glasgow on July 5 1823 with Dr George Birkbeck as its first president. Previous institutions, such as the Anderson Institute, which became the Glasgow Mechanics' Institute, and the Edinburgh School of Arts, had provided sound foundations through offering lectures, the use of a library and reading rooms. Glasgow is generally identified as the first mechanics' institute having this as its title.

Hudson (1969: 43), writing in 1850, states that in the first few months of operation the Glasgow Institute had around 600 members and the library, which had originated from the Andersonian Institution, had over 1000 books. Courses were offered in natural history, mathematics and anatomy. In the same month of July 1823, the Liverpool Mechanics' Institute and Apprentices' Library was opened which Kelly (1957: 75) believes to have been influenced by the New York Apprentices' Library which had been opened in 1820. This is not surprising as Liverpool had transatlantic connections with New York (see also Chapter 10, this volume).

In December 1823, both the Sheffield Mechanics' Institute and Apprentices' Library (not to be confused with the town's later Mechanics' Institute which opened in 1832), and the London Mechanics' Institution, were opened. The previous month saw the opening of the Kilmarnock Mechanics' Institute south of Glasgow, supported by John Steel, a former lecturer at Glasgow, and Greenock, close to the Port of Glasgow (Kelly 1957: 75). With the opening of the London Mechanics' Institution under Birkbeck's direction, Steel was one of a handful of men who are seen as the instigators of what was to become the Mechanics' Institute Movement, which spread across Britain, albeit unevenly.

Support for mechanics' institutes

From the geographical spread of the early mechanics' institutes, their origins developed through local needs and circumstances. Many were patronised by the first and second generation of industrialists, who themselves were from humble beginnings and mostly closely connected with expanding trade during the Industrial Revolution. As well as manufacturers, merchants and bankers also supported institutes such as those at Huddersfield, Leeds, Liverpool and Manchester (Kelly 1957). The banking family of John Smith Wright, for example, supported the foundation of the Nottingham Mechanics' Institute in 1837 (Green 1887) while Joseph Huntley and George Palmer, biscuit makers, supported the Newbury Institute in Berkshire (Thomas 1979). George Stephenson, engineer, supported the one at Newcastle; Josiah Wedgwood, pottery manufacturer and son of the founder, did the same at Hanley in Staffordshire, while Marc Isambard Brunel, engineer, supported the institutes at Rotherhithe and Bermondsey in London. Meanwhile, Benjamin Heywood, banker, was a passionate supporter of the Manchester Mechanics Institute, Joseph Strutt, a textile manufacturer, formed the Derby Institute and Charles Hindley, cotton spinner manufacturer, was behind the Ashton-under-Lyne Mechanics' Institute (Kelly 1957: 213).

Benjamin Gott, a wealthy woollen manufacturer in Leeds, contributed funds to the town's institute, opened in 1824, and was its first president (Garner and Jenkins 1984). These industrialists, bankers and engineers seem to have identified the potential for supporting the Movement.

It has been assumed that support came from the Radicals and Whigs, with opposition from the Church of England, the gentry and Tories. However, it is not quite so clear-cut as that. There were many instances where the 'landed gentry' had little interest in adult working-class education, and many looked upon institutes and similar as centres of political discontent and believed that educating the gentling masses might result in political revolution. However, Turner (1968) has identified that amongst the institutes founded in the Midlands, many were supported by the gentry. For example, the Duke of Sutherland was patron of the Hanley Institute and Lord Leigh was patron of both Birmingham and Coventry Institutes. Others supporting the Midland institutes included the Earl of Stamford and Viscounts Ingestre and Sandon, as well as Lords Lyttelton, Northwick, Sandys and Waterpark. In the case of the Yorkshire Union, Lords Fitzwilliam and Morpeth, Viscount Goderich, the Earls of Arundel and Ripon and the Dukes of Devonshire and Newcastle all supported the Movement. Some were Whigs, others Tories, others crossed over between the two parties. Joseph Strutt's nephew Edward Strutt was first Baron Belper and supported the family with establishing the Derby Institute. The sixth Duke of Devonshire was one-time President of the Institute (Chadwick 1975). The Devonshire family also gave land or the use of rent-free buildings for some of the institutes located in the Yorkshire Dales (*Annual Reports of the Yorkshire Union of Mechanics' Institutes 1839–1900*). Sir John William Ramsden, the largest landowner in Huddersfield, provided financial support for the town's Mechanics' Institute. As Turner (1968: 67) observes, 'substantial support was given by various members of the peerage ... and because these men were generally very powerful in the locality, their support was particularly valuable'.

With regard to the role of religious influence, like the landed gentry, there was a mixed reception. The Anglican vicar of Woodlesford and Oulton in Yorkshire, after which the Institute of the same name was founded in 1847, stated that 'our little Institute is a spark which, if encouraged will burst out into a prodigious conflagration and destroy every desire for labour and order' (Turner 1968: 65). However, at nearby Leeds the Institute committee members included an Anglican clergyman, and Benjamin Gott, the first president, was an Anglican, while another manufacturer on the committee, James Holdforth, was a Roman Catholic (Garner and Jenkins 1984: 142). In the Midlands, sixteen Anglican clergymen supported the foundation of the Birmingham, Coventry, Hanley and West Bromwich Institutes in the Midlands, with two Roman Catholics priests, along with twenty-one Dissenters who were made up of 11 Independents, five Unitarians, three Baptists, a Presbyterian and a Wesleyan (Turner 1968). Chadwick (1975) noted that The Strutt family, founders and presidents of the Derby Institute, were Nonconformists, and Inkster (1976) states that support for an Institute at Rotherham near Sheffield came from Quakers, Unitarians and a Roman Catholic. Nationally, it was the Dissenters, particularly Unitarians, who were identified most closely with the Mechanics' Institute Movement through their theological beliefs and their support for social reform including providing education for all – boys and girls, men and women.

Some Members of Parliament of all political persuasions did not support the Movement, while others supported it. Turner (1968) identified that across the three Midland counties there were thirteen Whigs, nine Tories and five Radicals who were very much involved with their local institutes. Tories supported the Hanley and Stafford Institutes, and both the Tories and the Whigs supported Birmingham, Longton, Newcastle in Staffordshire, Stourbridge and Walsall. The Strutt family, founders of the Derby Institute, were Whigs, while at Leeds, over the period between 1823 and 1835, there were eight Tories, thirty-four Whigs, one Radical and fourteen 'unknown' (Garner and Jenkins 1984: 142). The Devonshire Family, who supported several insititutes in Derbyshire and Yorkshire, were Whigs.

As previously mentioned in Chapter 1 of this volume, mechanics' institutes were emerging during a period of social reform and political agitation. The Committees of mechanics' institutes often listed as their first rule discourage-ment of political or sectarian activities that would bring them or their members into disrepute. They did not want to be seen by the political class as contributing to social unrest, especially as members from the operative class may well have had sympathies in support of political change. On the other hand, many Radicals, anxious to support the education of the masses and seen by the Tories as want-ing to change the status quo of society, were frustrated that politics, economics and religion were not part of the curriculum, and many left the Movement. As Inkster (1976: 288) notes, the promoters of the institutes were often identified for their more than passing interest in science, 'generally reformist in politics and dissenting in religion'.

The early mechanics' institutes also had support from their local newspapers, providing of course they had political sympathies towards supporting them. For example, the *Liverpool Mercury*, the *Sheffield Independent*, the *Leeds Mercury* and the *Dumfriesshire and Galloway Courier* all supported their local institutes. Their contribution in supporting the initiative should not be underestimated. The *Birmingham Journal* was the voice of a group of manufacturers in the town who supported mechanics' institutes but were concerned that they could become centres of discontent and often shared caution with their readership that insti-tutes could become centres for political agitators. Some also published the annual reports of their local institute, and Edward Baines Senior, the *Leeds Mercury* pro-prietor, not only published the Leeds Mechanics' Institute annual reports but also reports for those institutes that were members of the Yorkshire Union. In the case of Leeds, it was the Editor of the *Leeds Mercury*, Edward Baines, and his son, Edward Junior[1], who were the leading figures in establishing the Institute, and had published a suggestion initially to establish a library for 'mechanics and artisans, chiefly of the labouring classes' for an annual fee of five shillings (Garner and Jenkins 1984: 140). An article in the *Leeds Mercury* of 20 December 1823, thought to have been written by Edward Baines Senior, stated:

I entirely agree with you as to the beneficial effects likely to flow from the proposed School of Arts. Go on, and prosper, in God's name! Let the rich spare something from their abundance for institutions like this: let them disseminate knowledge amongst the ignorant, and allure the working classes into good habits; by improving the intellects and refining the tastes of the poor, you promote both industry and morals, both their present and their best interests (ibid.).

Often a School of Arts was set up as part of an institute. Baines thought the Leeds Institute should be managed by the middle-class Committee so they had social control and could safeguard the morals of artisan members who might otherwise be spending their time in the public houses. Some newspapers, such as the *Birmingham Journal*, whilst supporting institutes in attracting artisans to learn about science, were more enthused about preventing them from entering public houses.

This underlining concern was also taken up by the Temperance Movement, which supported mechanics' institutes and other educational voluntary establishments. John Finch, an iron merchant in Liverpool and supporter of the Temperance Movement, noted in 1833 that beer-house keepers and publicans offer the use of their premises for benefit societies, money clubs and friendly societies, where beer cost no more than tea and was safer to drink than water or milk. There was also warmth, light, newspapers, books and proper lavatories. In Nottingham there were libraries in public houses and it was noted in Pudsey, in the West Riding of Yorkshire, that the only places for spending free time when not at home was the church, chapel or local public house, and the latter was open more hours every day and provided various kinds of amusement.

The middle class therefore began to preach what Cunningham (1980: 33) calls 'rational recreation' to the labouring population. The Temperance Movement, like voluntary education organisations such as mechanics' institutes, had its origins in middle-class, mainly Protestant culture, and amongst social reformers and philanthropists. With industrialisation and the consequential spread of urbanisation and poverty, these groups were concerned that lower orders should be supported with self-improvement, self-control and recreation of a cultural kind, as opposed to unruly pastimes such as drinking alcohol in the public houses. The temperance and education societies shared a belief that such middle-class values should be passed on to the lower groups in society so as to 'civilise' them, to provide social control over potentially troublesome groups fuelled by alcohol.

Like working-class radicals and the Chartists, the Temperance Movement identified the potential of mechanics' institutes and public libraries as sources of a new leisure: the leisure of learning. Between 1833 and 1872, for example, Harrison (1971: 174) identifies that of 100 teetotallers, all of whom were supporting prohibition, 41 supported anti-slavery, 40 were members of the Anti–Corn Law League, 22 were involved with their local mechanics' institutes, 12 were supporting voluntary education, six were supporting the free press and six the establishment of public libraries. The Temperance Movement was particularly popular amongst Quakers in promoting peace and penal reform as well as education.

Both mechanics' institutes and temperance societies appealed to middle and lower-middle classes as well as the respectable working classes who, having become both educated and sober, could gain useful knowledge which would increase their potential for better employment opportunities, prosperity and better social standing in society. Both mechanics' institutes and temperance societies were also independent of any political or religious influences and therefore safe places to be associated with.

Thus, the temperance society was keen to support the Movement and distract those who poorly behaved or spent much of their time in the public houses, through offering them mental improvement and leisure activities. Sir Edward Baines Junior, the Whig editor of the *Leeds Mercury*, summed up this feeling when describing Holbeck near Leeds, where an institute was being proposed: '[T]he great part of the population was composed of thoughtless youths; some illiterate as savages ... drunkenness among young men was a prevailing vice, street brawling was a daily occurrence' (ibid.). However, Thomas (1979: 70–71) has identified from the Leighton Buzzard Working Men's Mutual Society Minutes that,

> the pub was anathema to most reformers, its ubiquity and popularity in working-class life made it an inescapable model, and institutes tried to reproduce its social and physical warmth (the fact that a fire was kept was emphasized in the institutes promotional activities) and were also to compete with pub creations such as draughts, chess and railway excursions.

Joseph Liversey, a former child loom operator and later a cheesemonger in adult life, formed Preston Temperance Society, which was supported by the *Preston Chronicle*. He also supported the Preston Mechanics' Institute, the forerunner of the later Harris Institute (Harrison 1971). The foundation of the Rotherham Mechanics' Institute in 1838 is one example of many that were established across the country by the local branch of the temperance society, 'in an effort to combat the region's local reputation of habitual drunkenness'. A former vicarage was purchased by the Society for the purpose of delivering lectures, with a library for 1,500 books 'and a laboratory for scientific experiments' (ibid.: 178). Edward Baines, President of the Leeds Mechanics' Institute, was one of the early supporters of the Rotherham Mechanics' Institute. In 1843 the Temperance Society donated a further £400 for 'books, equipment and class preparation' to the Institute (*First Annual Report of the West Riding of the Yorkshire Union* 1838: 126–7). Later, the Institute moved into its own premises (Inkster 1976). Also in the West Riding at Huddersfield, the first president and benefactor of the reopened town's Mechanics' Institute in 1841, Frederic Schwann, was also president of the local temperance society for thirty years (*Huddersfield Examiner* 25 April 1882). Working men's clubs, several of which are known to have been members of the Yorkshire Union of Mechanics' Institutes, became active from the 1860s, though not offering beer and spirits. Instead, they provided leisure activities and sometimes reading groups and libraries, the latter being supported by the union.

In the North East, the establishment of the Newcastle Literary, Scientific and Mechanical Institute was seen by the middle class as a positive contribution because it tended 'to improve the morals of a rising generation, to guard them from the seductive pleasures, from drunkenness and dissipation' (*Tyne Mercury* 2 March 1824). The Berkhamsted Mechanics' Institute in Hertford was supported by its local temperance society and amongst its members was George Cruikshank, the political cartoonist. Temperance societies sometimes set up their own institutions such as that at Reading in Berkshire, where the local society established a reading room, library and evening college (Thomas 1979). The Temperance Movement was particularly active in the Midlands and North, with much support in Lancashire, the West Riding of Yorkshire and North East England (Harrison 1971). These were areas where the Mechanics' Institute Unions were particularly active.

The beginnings of the Mechanics' Institute Movement

The Glasgow Mechanics' Institute included as its main aim not to rely on 'wealthy citizens' for financial support in return for decision-making and for managing it. Instead, contributions came from them to fund it, as well as from subscriptions from members themselves (Hudson 1969). On this basis, the Institute set the tone on which the Movement spread across the whole of Britain, and over the next 27 years mechanics' institutes increased rapidly by number, if not membership; by 1850 there were around 600 similar institutions across the British Isles (Hudson 1969).

During 1824 there were at least three mechanics' institutes established in Scotland: Aberdeen, Hawick and Dundee. In England, nine were opened: Kendal, Lancaster and Manchester in Lancashire, Eyam in Derbyshire, Leeds in Yorkshire, Alnwick and Newcastle in Northumberland and, in the south, institutes were formed at Bury St Edmunds and Ipswich in Suffolk. The first one established in Wales was the Bridgend Mechanics' Institute in Glamorgan.[2] Already, the Movement was not just associated with the growing industrial towns such as Manchester, but also the more rural areas such as the former plague village and lead-mining settlement of Eyam.

Kelly (1957) believed 1825 to be the vintage year, during which 70 new institutes were established. In the London suburbs alone, there were at least six: Bermondsey, Camberwell, Deptford, Hackney, Hammersmith and Spitalfields. In Scotland, there was activity around the Forth–Clyde region and suburbs of Glasgow. There was also expansion in Lancashire with new institutes being established at Ashton-under-Lyne, Bolton, Liverpool, which merged with the School of Arts, Warrington and Wigan (Hudson 1969). Others have been identified: in Cheshire, institutes at Stalybridge and Stockport were established and in Yorkshire 13 were opened, including those at Bradford, Dewsbury, Halifax, Huddersfield,

Hull and Wakefield. Meanwhile, in the North East counties of Durham and Northumberland, mechanics' institutes were established at Darlington, South Shields, Stockton, Sunderland and Tynemouth. Others were established at ports, for example, at Bristol in Gloucestershire, Devonport, Plymouth and Portsmouth in Devon, and in the North at Whitehaven in Cumberland.

Stagnation and decline: 1826–1850

Despite such a quick rise and supporting foundation on which mechanics' institutes had been established, in less than four years, by 1826, there was already some indication that at best it was stalling or, worse still, declining. Indeed, Kelly (1957) refers to this period as one of sudden collapse. Three examples are Bradford, which closed and only reopened in 1832, Huddersfield, which closed in 1826 and reopened in 1841 and Birmingham, opened in 1825, closed in 1840 and later reopened. Only 13 new institutes, among them Mansfield, Nottinghamshire, were opened in 1831 (Briscoe 1954), compared to 69 the year before, and no more than nine opened each subsequent year from 1832. Initial explanation could be due to the fact that most towns that were able to sustain an institute had now got one. However, on closer examination, many that had been open previously had now closed, such as those in the London suburbs and small towns. In the North, Bury, Darlington, Huddersfield Rotherham and Skipton all closed. Those that were still operating were doing so on a much smaller membership such as those at Aberdeen, Ashton-under-Lyne, Portsmouth and York. Even the larger ones at Liverpool, London and Manchester saw their numbers fall (Kelly 1957: 223).

The primary cause was the economic depression of 1826 that resulted in both unemployment and falling wages, meaning that working men could no longer afford to pay the quarterly membership fees. Huddersfield Institute closed sometime in 1826 as the bank withdrew its loan as it needed to find higher returns from other ventures. There was also a lot of uncertainty in the textile districts of the North, particularly with strikes and rioting due to the severe cut in the wages of spinners. The economic uncertainty also resulted in less financial support from patrons, many of whom worked in manufacturing and trade, and instead had to put what money they had into their businesses rather than the local institute, which provided no financial return. There was a further downturn in the economy during the 1830s.

Second, the quality of lectures varied substantially, with the larger institutes being able to attract the very best, often paying five guineas (£5.50) a time. With the reduction in income, this was no longer sustainable. The 'whiz-bang' of new scientific developments being publicised in classes and lectures where not that frequent, and the theoretical underpinning was often hard to follow by the membership. They were after all artisans, wanting to learn about work-related subjects rather than calculating formulae. Even by the 1840s things had not improved. The *Second Annual Report of the Yorkshire Union of Mechanics' Institutes* (1840: 7–8) put it succinctly when identifying the fall in membership was 'unquestionably to be found in the absence of an early and sound moral training for the mass of our operative population' and the need for elementary education. In 1844, the Yorkshire Union identified the problem by stating 'we have endeavoured to form colleges before we had schools' (*Sixth Annual Report of the Yorkshire Union of Mechanics' Institutes* 1844: 7).

End of the Movement?

Between the early 1830s and 1840s, the number of new or reopened institutes never resulted in a momentum of activity such as that seen before 1826. There was some growth in the London suburbs: Croydon, Greenwich, Tottenham and Woolwich to name but four and adding to others that had reopened. There was also further activity in Central Scotland. In Lancashire, Liverpool had three, and Manchester and Salford had six each, as well as several new ones being established in East Lancashire. Others reopened, such as Bury and Wigan. Yorkshire, too, saw some expansion of institutes with several new ones around Bradford, Halifax, Huddersfield (the town's own institute only reopened in 1841), Leeds and Wakefield, and in the south of the county around Sheffield with Barnsley reopened for the third time in 1837 (Kelly 1857: 230–31). Sheffield Mechanics' Institute was only founded in 1832, possibly due to the previous success of the town's Mechanics' and Apprentices' Library, which had been very proactive, particularly in accepting children who had previously attended local schools and who had access to its library (Salt 1966).

There was also a scattering across the rest of the country, for example, in the Midlands, Mansfield had an institute in 1831, Nottingham Mechanics' Institute was formed in 1837, although there had been a Mechanics' and Apprentices' Library established in the town in 1824, and in nearby Newark in 1836 (Briscoe 1954), East Anglia, the southern counties and stretching into South Wales and the West Country, and, by 1850, there were over 700 similar institutes (Hudson 1969).

Research carried out by Thomas (1979) found that while only 3 per cent of mechanics' institutes were found in the Home Counties in 1851, they nevertheless supported their local population.[3] Kelly (1957) believed that there were only 30 such institutes in the Home Counties but Thomas (1979) states that the number was nearer 90 at least. While the number of institutes were not on the scale of those found in the industrial North and Midlands, as the Home Counties were predominately agricultural, 30 per cent of the population were working in small-scale industries such as cloth weaving, brewing, brick making, printing and ironmaking. Mechanics' Institutes were established at Banbury (1836), Bedford, Berkhamstead, Candleford, Hitchin (1835), Luton, Newbury, Reading (1825), Royston, Witney and Wolverton, the latter for the workers of the railway workshops. Figure 2.1 shows the distribution of the Movement in 1850, taken from the lists for each county produced by Hudson (1969).

Not surprisingly, the industrial areas of Central Scotland, South and East Lancashire, the West Riding of Yorkshire (the county was divided into three 'Ridings' from the Saxon term meaning 'one third', made up of West Riding, East

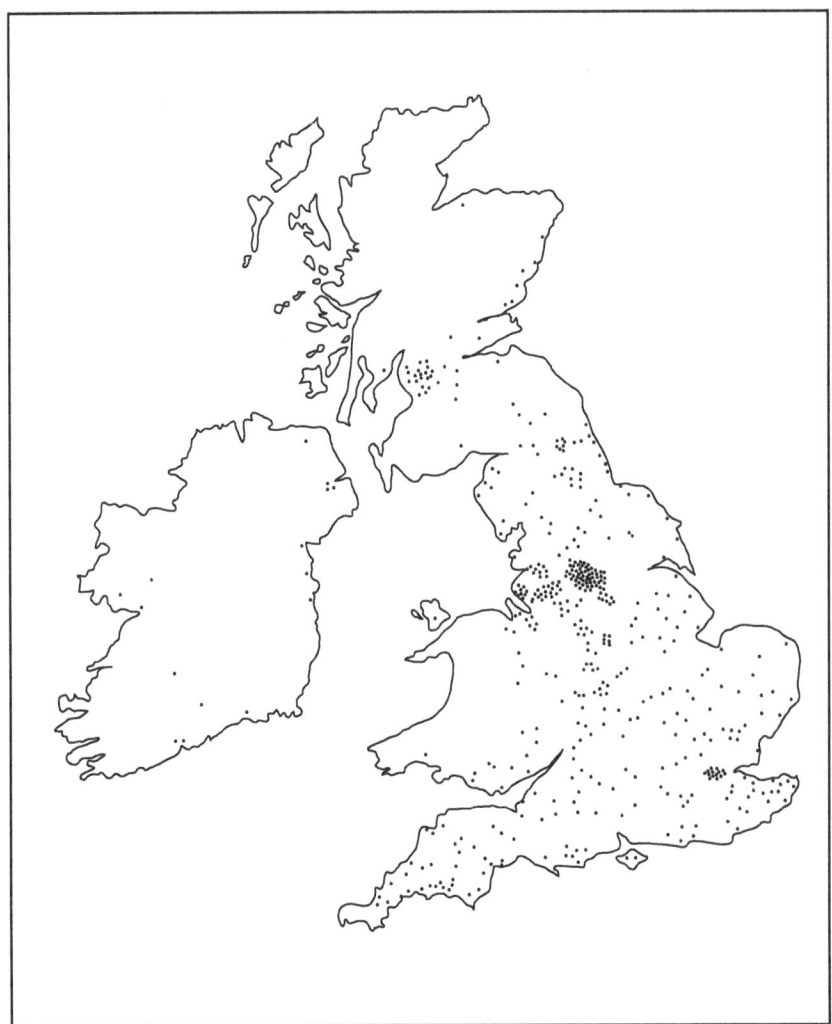

Figure 2.1 The distribution of mechanics' institutes across Britain by 1850.

Source: Data from Hudson 1969.

Riding and North Riding) and the West Midlands, particularly in Staffordshire, show the popularity that mechanics' institutes had in industrialising towns. However, the rest of the country also had a scattering. Rutland, the smallest county, not surprisingly, did not have an institute, but was close to Leicester in any case. Those in Wales were mainly on the coalfields in the south. Wales had a strong Sunday School Movement, particularly in the North, which would have offered education to both children and adults.

With regard to Ireland, the majority of the country was rural with little industrialisation outside the towns of Belfast and Dublin. Indeed, Kelly (1957: 252) was surprised that the Mechanics' Institute Movement in Ireland was as active as it was, given

> that in a country riven by political and sectarian animosities, with few large towns and little developed industry, and with a poverty-stricken population, a movement directed toward the scientific education of the working people should gain a foothold at all.

Duffy (1990) noted that, whereas in other parts of Britain the intention was to educate the workforce in the emerging industrial society, the hope was in Ireland that technical education would inspire its workforce to bring about their own industrial revolution.

Neswald (2008: 212) states that the Mechanics' Institute Movement in Ireland 'received an enthusiastic response', with the first institute being established in Dublin in 1824 and followed in 1825 by others being founded in Belfast, Cork, Limerick, Waterford, Armagh and Ennis. Further institutes were formed at Galway in 1826, at Nenagh and Portaferry in 1828 and Londonderry in 1829. Between 1824 and 1860, over 30 mechanics' institutes and similar voluntary organisations 'for the education of adult and apprentice tradesmen and artisans' were established in Ireland.

A new age, a new start

The Mechanics' Institutes Movement did continue to develop after 1850, either through existing institutes that had reopened and adapted to changing times, or through new ones. In Nottinghamshire, for example, the Worksop Mechanics' Institute was opened in 1852 (Briscoe 1954). Meanwhile in Birmingham, a second institute, the Birmingham and Midland Institute, was formed in 1854, replacing its predecessor opened in 1825 and closed in 1840, highlighting that even in larger industrial areas the Movement had struggled. In the case of Yorkshire, particularly in the more rural areas of the East and North Ridings, new institutes were being established, such as those at Bedale (1850), Great Ayton, Guisborough (1850), Helmsley (1850), Hutton Rudby (1851) and Howden-le-Wear (1876). Those new institutes established in the industrial West Riding of Yorkshire included Boston Spa (1850), Calverly (1850) and Oakworth (1850) (*Twelfth Annual Report of the Yorkshire Union of Mechanics' Institutes* 1850).

What was happening in Yorkshire was also happening elsewhere; that is, institutes being established in 1850 or 1851 in urban or rural areas, for example, Coniston, Denton, Droylsden, Slaidburn and Wavertree in Lancashire, Sandbach and Styal in Cheshire, Falmouth, Penryn and Penzance in Cornwall, Dronfield in Derbyshire, Leadgate and West Hartlepool in County Durham and in Kent, Bexley and Sevenoaks (Walker 2010). In all, there were around 35 new institutes established between 1850 and 1851. These can be added to the 700 or so already established and operating up and down the country.

From 1853, with the establishing of the Science and Art Department in that year, many established institutes benefited from the funds available to them, including those at Mansfield and Newark in Nottinghamshire (Briscoe 1954). Using data available from the Yorkshire Union of Mechanics' Institutes, which also included member institutes in Lancashire, Westmorland and County Durham, several new institutes were founded which were specialist ones such as the Darlington Railway Institute established in 1868, Saltaire in 1873 at Sir Titus Salt's textile model village near Bradford, Esh Miners' Institute and Skinningrove Miners' Institute, both established in 1877. Other more general institutes were founded in rural areas, such as the ones at Earby in 1859 and Gisburn in 1877. Others were founded in urban areas such as at Darlington's Albert Hill in 1870 and at Headingley, Leeds, in 1877. Some towns that had never previously had a mechanics' institute included Pontefract in 1870 and Beverley, Cleckheaton and Harrogate in 1873 (*Thirty-Fifth Annual Report of the Yorkshire Union of Mechanics' Institutes* 1873). From the late 1870s until the turn of the century most Yorkshire Union institutions were village libraries, some of which offered evening classes as well as a circulation of books provided by the Yorkshire Union (Walker 2013). Similar developments were taking place across the country, including those members the Lancashire and Cheshire Union.

The revitalised Society of Arts, following the 1851 Exhibition, substantially revitalised the Movement (Walker 2014). Many institutes also became centres for the City and Guilds of London Institute, which took over responsibilities for technical subjects examinations from the Society of Arts. There were also the University Extension Schemes offered by, among others, Oxford, Cambridge and Durham, which were first introduced in the 1870s, at many of the larger institutes. These developments will be discussed in more detail in later chapters.

Extensive research had identified that several mechanics' institutes that opened in the 1820s and 1830s closed and then reopened. The Yorkshire Union alone has several examples such as Barnsley, which was opened in 1831 and reopened in 1848, Bradford, opened in 1825 and reopened in 1832, and Huddersfield, which opened in 1825 and reopened twice, the first time in 1838 and again in 1842. This also happened at the Skipton Institute, which opened in 1825, and reopened in 1845 and again in 1848. Such examples have reinforced education historians' and other academics' views that the Movement overall was a failure.

In the case of the Yorkshire Union of Mechanics' Institutes, Laurent (1984: 589) identified that there was 'continued vigour in the latter part of the nineteenth century'. Table 2.1 provides the number of institutes that were members of the Yorkshire Union and therefore does not include all those that were established in Yorkshire. For example, Huddersfield Mechanics' Institute was founded in 1825 but did not become a member of the union until 1844, three years after it was reopened, and Sheffield, which was opened in 1832, did not join until 1838 (Walker 2010).

Blake (1859) listed the number of Yorkshire Union Institutes between 1846 and 1859 but identified that out of the 151 Yorkshire Institutes listed by Hudson in 1851 (1969), only 109 were members. Laurent (1984) provides the number

Table 2.1 Number of Yorkshire Union Mechanics' Institutes and
membership 1846–1890

Year	Total number of institutes	Total number of members
1830	9	Not known
1846	48	9,034
1847	63	12,000
1848	79	14,105
1849	86	15,860
1850	109	18,516
1851	117	19,293
1852	123	19,043
1853	127	19,362
1854	127	20,305
1855	133	21,080
1856	130	21,297
1857	130	20,960
1858	128	22,600
1859	138	24,600
1868	122	23,000
1879	245	46,000
1890	276	60,000

Sources: Blake 1859: 335; Laurent 1984: 585.

of Yorkshire Union Institutes from 1869 to 1890, and some were located in
other counties such as County Durham and Lancashire. In total, between 1824
and 1900, there were 633 institutes, reading rooms, village libraries and working
men's clubs that were offering some kind of education that were also members
of the Yorkshire Union. These figures do not take into account perhaps smaller
institutes that could not afford the union's annual membership subscriptions.

Hemming (1977: 18) believes that around a third of institute membership in
1850 was concentrated in Cheshire, Lancashire and Yorkshire. Within the West
Riding of Yorkshire alone, there were at least 130 mechanics' institutes with a
combined annual average of 20,000 members between 1850 and 1868. In the
case of the Lancashire and Cheshire Union, in 1868 there were around 100 institutes with a total average of 24,000 members. Between 1851 and 1900, the four
Yorkshire Union institutes of Bradford, Huddersfield and Keighley averaged over
1000 members per year and Leeds in excess 2000 over the same period. In the
case of the Lancashire and Cheshire Union, Burnley, Manchester, Preston and
Stockport had similar averages. Across both the Lancashire and Cheshire and the
Yorkshire Unions, 10 per cent were made up of females (see Chapter 6 of this
volume).

Conclusion

From the historical overview presented here, there is ample evidence to support
the view that the Mechanics' Institute Movement not only survived beyond the

1850s but continued to expand until the end of the nineteenth century. The Movement spread from Glasgow across the whole of Britain, including Ireland, not only in the growing industrial heartlands but also in semi-rural and rural areas of the country. While Birkbeck and Brougham can be identified as the founding fathers, unlike other Victorian movements the Mechanics' Institutes Movement was not supported by one group in society. Apart from the self-made industrialists who were enthusiastic in supporting them, aristocrats, landed gentry and the professional classes were also supportive of mechanics' institutes. Politically, Liberals, Radicals, Tories and Whigs all had both critics and supporters, depending on their own leanings in politics, as was also true for religious groups, particularly Anglicans, who gave them a mixed reception. On the other hand Dissenters, including Roman Catholics, Unitarians, and Quakers not surprisingly, supported them. The Temperance Society through its local branches supported the establishment of mechanics' institutes, providing funds for books and equipment. In other words the Movement was supported across the class, politics and religious divides. This chapter has successfully argued that the Mechanics' Institute Movement did continue beyond 1850 until the end of the century. However, what is not apparent at this stage is to what extent mechanics' institutes were an educational success. There is much evidence that they seem to have often opened and closed, some several times. Why? Did they support the class for whom they were established and what, if any, impact did they have on the early twentieth-century state-funded adult and further education? The answers to these and other questions will be covered in the following chapters.

Notes

1 Edward Baines was a journalist, politician, Congregationalist and educationalist who was born in Leeds in 1800. He was a supporter of educational voluntarism for learners of all ages until later in life when he supported the need for state involvement. A founder of the Leeds Literary and Philosophical Society 'for his own class', he did also support working-class adult education after visiting Dr George Birkbeck's Mechanics' Institute in London in 1824. Baines was the instigator in establishing several institutes in Yorkshire during the 1820s and bringing them together to form the West Riding Union and in 1838 the Yorkshire Union of Mechanics' Institutes. He was President for forty years. Baines also supported the Elementary Education Act of 1870 and established the Yorkshire College of Science, which would later become Victoria University and then Leeds University.

2 This study makes reference to the old counties of Britain, such as the three Ridings of Yorkshire, Cumberland and Westmorland as they were during the period of study, prior to the renaming and boundary changes under the Local Government Act of 1972.

3 There is no exact definition of the term 'home counties' but they are often referred to as those counties of England that surround London and include Berkshire, Buckinghamshire, Essex, Hertfordshire, Kent and Surrey. Other counties more distant from London, such as Bedfordshire, Cambridgeshire, Dorset, Hampshire, Oxfordshire and Sussex are also sometimes included in the list due to their close proximity to the capital and their connection to the London regional economy.

References

Blake, B. (1859) 'The Mechanics' Institutes of Yorkshire', *Transactions of the National Association for the Promotion of Social Science*, 355–364.

Briscoe, H. K. (1954) 'Nottinghamshire Mechanics' Institutes in the nineteenth century', *The Vocational Aspect of Education* Vol. 6, No. 13, 151–162.

Chadwick, A. F. (1975) 'The Derby Mechanics' Institute 1825–1880', *The Vocational Aspect of Education* Vol. 27, No. 68, 103–105.

Cunningham, H. (1980) *Leisure in the Industrial Revolution circa 1780–circa 1880* (London: Croom Helm).

Dick, M. M. (2008) 'Discourses for the new industrial world: Industrialisation and the education of the public in late eighteenth-century Britain', *History of Education: Journal of the History of Education Society* Vol. 37, No. 4, 567–584.

Duffy, S. (1990) 'Treasures open to the wise: A survey of early Mechanics' Institutes and similar organisations', *Irish Labour History Society* Vol. 15, 39–47.

Garner, A. D., and Jenkins, E. W. (1984) 'The English Mechanics' Institutes: The case of Leeds 1824–1842', *History of Education: Journal of the History of Education Society* Vol. 13, No. 2, 139–152.

Green, J. A. H. (1887) *History of the Nottingham Mechanics' Institution, 1837–1887* (Nottingham: Stevenson, Bailey and Smith).

Harrison, B. (1971) *Drink and the Victorians, The Temperance Question in England 1815–1872* (London: Faber and Faber).

Hemming, J. P. (1977) 'Mechanics' Institutes in Lancashire and Yorkshire Textile Districts from 1850', *Journal of Educational Administration and History* Vol. 9, No. 1, 18–31.

Huddersfield Examiner (25 April 1882) 'Temperance Society' (author unknown).

Hudson, J. W. (1851) *The History of Adult Education in which is comprised a Full and Complete History of the Mechanics' and Literacy Institutions* (Reprint. London: Woburn 1969).

Inkster, I. (1976) 'The social context of an educational movement: A revisionist approach to the English Mechanics' Institutes, 1820–1850', *Oxford Review of Education* Vol. 2, No. 3, 277–307.

Kelly, T. (1957) *George Birkbeck: Pioneer of Adult Education* (Liverpool: Liverpool University Press).

Laurent, J. (1984) 'Science, society and politics in late nineteenth century England: A further look at Mechanics' Institutes', *Social Studies of Science* Vol. 14, 585–619.

Neswald, E. (2008) 'The benefits of a Mechanics' Institute and the blessing of temperance: Science and temperance in 1840s Ireland', *Social History of Alcohol and Drugs* Vol. 22, No. 2, 20–227.

Salt, J. (1966) 'The creation of the Sheffield Mechanics' Institute', *Vocational Aspect of Education* Vol. XVIII, No. 40, 143–150.

Smiles, Samuel (1865) *Lives of Boulton and Watt* (Philadelphia, U.S.A: J. B. Lipincott).

Thomas, R. A. (1979) 'The mechanics' institutes of the Home Counties, circa 1825–1870 Part One', *The Vocational Aspect of Education* Vol. 31 No. 79, 67–72.

Turner, C. M. (1968) 'Political, religious, and occupational support in the early Mechanics' Institutes', *The Vocational Aspect of Education* Vol. XX, No. 45, 65–70.

Tyne Mercury (2 March 1824) 'Aims of the Temperance Society' (author unknown).

Walker, M. (2010) '*Solid and practical education within reach of the humblest means: The growth and development of the Yorkshire Union of Mechanics' Institutes 1838–1891*, unpublished PhD thesis (Huddersfield: University of Huddersfield).

Walker, M. (2013) 'For the last many years in England everybody has been educating the people, but they have forgotten to find them any books: The Mechanics' Institutes Library Movement and its contribution to working-class adult education during the nineteenth century', *Library and Information History* Vol. 29, No. 4, 272 286.

Walker, M. (2014) 'Earnest students anxious to acquire a practical knowledge suited to the trade of the district: The growth and development of the Mechanics' Institute Movement with particular reference to Huddersfield 1824–1890', *Journal of Educational Administration and History* Vol. 46, No. 1, 38–56.

West Riding of Yorkshire Union of Mechanics' Institutes (1838) *First Annual Report*, 126–127.

Yorkshire Union of Mechanics' Institutes (1840) *Second Annual Report*.

Yorkshire Union of Mechanics' Institutes (1844) *Sixth Annual Report*.

Yorkshire Union of Mechanics' Institutes (1850) *Twelfth Annual Report*.

Yorkshire Union of Mechanics' Institutes (1873) *Thirty-Fifth Annual Report*.

3 Scientific education offered up to 1850

Introduction

This chapter concentrates on what education and training the mechanics' institutes offered. It will also attempt to identify what evidence historians and educationalists have put forward to support their views that the Movement was in decline by 1850. 'Education and training' is a relatively recent term which covers post–school qualifications relating to curricula associated with academic subjects such as 'A' Levels, vocational programmes and apprenticeships. Alongside these are foundation courses offered in colleges as well as basic skills in literacy and numeracy programmes. Previously, courses relating to work-based employment were referred to as technical education, and foundation and basic skills as elementary education.

The growing interest in science can be traced back to the seventeenth century and by the eighteenth century there was a need for its practical application, particularly with regard to the emerging industrial developments that were beginning to rely on technology to support their expansion. As Peters (1963) noted, during the eighteenth century public lectures on natural philosophy or experimental philosophy were given by itinerant lecturers and financed by their middle-class audiences. In 1819, John Griscom, Professor of Chemistry at the New York Institution, reported that he had attended a public science lecture given at York in England (Inkster 1980: 80).

> There is scarcely a town of considerable note in Great Britain, which is not sometimes visited by these travelling lecturers, who, by means of portable apparatus, and a facility in communicating instruction, import the benefits of useful knowledge to hundreds and thousands who might otherwise remain destitute of its advantages (Griscom 1824: 198).

Lectures were given across the country but particularly in the North. With the establishment of mechanics' institutes in the growing towns, the tradition of itinerant lecturers continued as a way of spreading scientific knowledge to the populace; for example, several visited Hull, Newcastle, Liverpool, Sheffield, Derby, Lancaster, Birmingham and Warrington on a wide variety of topics (Inkster 1980: 91). They were in essence the first lectures on practical scientific knowledge.

In Britain, two important organisations were established in support of the practical application of science: in 1755, the Society of Arts for the Encouragement of Arts, Manufacture and Commerce (hereafter referred to as the Society of Arts), and in 1810 the Royal Institution was established as a public body responsible for the promotion of chemical science through experiments, lectures, mineral research for industry and useful knowledge in general (Cotgrove 1958). This was in contrast to the classical sciences offered through traditional middle-class schooling and the universities. There seems to have been the expertise, but to what extent this was reaching the masses is unknown.

Curriculum developments 1824–1850 in the mechanics' institutes

The overall aim of mechanics' institutes was to provide access to scientific knowledge and specifically to make their members efficient industrially through 'useful knowledge' as presented by their middle-class promoters who were wishing to attract mechanics (Cotgrove 1958). Barnard (1966: 180) argues that mechanics' institutes 'tended to fail because they provided not so much vocational training for working men, as general educational and social facilities for members of the professional and middle class'. Barnard's work has therefore reinforced the view, held by many historians and educationalists, that mechanics' institutes were not educationally successful in supporting the needs of the working classes.

It is certainly true that in many cases during the 1820s and 1830s, when mechanics' institutes were being established, most did offer scientific curricula on the assumption that this was what they should be offering to adults in support of technical education for industry. Several mechanics' institutes across the country, including the one at Sheffield, had highly reputable scientific classes and public lectures and, while they were open to everyone of whatever class, the majority of members were middle-class professionals who could relate to the high level of scientific knowledge being presented at lectures. The Institute offered natural philosophy lessons on a weekly basis between 1838 and 1841, supported with itinerant lecturers delivering public lectures in science. Lectures were given in phrenology, astronomy, mechanics, chemistry and electricity and were given by local men including a professional chemist, botanist, surgeon and a steel manufacturer (Inkster 1975).

Inkster (1975: 451) identified that many mechanics' institutes traced their origins back to establishing themselves specifically for science, among them Derby, Newcastle, Nottingham and Sheffield. Their scientific credibility is not being questioned; many institutes had leading scientists as visiting lecturers. However, the depth and content of the lectures did not attract the class for which they were established. At Newcastle, for example, there was an emphasis on the reading of papers on topics such as 'the proper measure of the force of a body in motion'. At the same institution classes were held in 'chemistry and higher mathematics', not practical and technical subjects relevant to a working-class population.

The main object of the Sheffield Mechanics' Institute had been 'to supply, at a cheap rate, education to the [labouring] classes of the community' and provide provision of 'solid and useful instruction in the various branches of Art and Science, more particularly in such as are connected with the Staple Manufacturers of the Town'. However, in reality 'the initial response of the working class to such middle class efforts was weak' (ibid.: 453). Occupations of members of the Sheffield Mechanics' Institute verify that in 1832 there were no members from the labouring classes (ibid.: 470). Yet as Salt (1966: 144) suggests, the initial support for the Institute had been 'to stimulate economic progress through education' and particularly as a result of the cutlery, edge tool and ornamental metalwork manufacturers who were identifying that there was overseas competition in the markets they had had the monopoly on. Although the Institute's original aims are commendable with regard to adult education for the masses, Sheffield had a coherent 'scientific community' of middle-class members which in itself must have discouraged the working class from partaking because of the level and content of lectures as well as being off-putting to those who were in reality 'outsiders'.

In the case of the Manchester Mechanics' Institute, there had been 'great public excitement' when, in 1825, the first lectures were delivered by the Rev. Andrew Wilson on mechanical philosophy and Richard Phillips, Fellow of the Royal Society[1], on chemistry, and 'yet only a few weeks later the enthusiasm had waned and the audience has been lost' (Tylecote 1974: 60–61). When analysed, it was noted by John Dalton that such high-level scientific knowledge could not be appreciated by those to whom it was offered because they lacked basic knowledge in the subject.[2] Dalton, one of the founders of the Manchester Mechanics' Institute, wrote to one of the other fellow founders, Benjamin Heywood, stating:

> It is highly necessary for the hearers previously to acquire a good knowledge of arithmetic and to be in some degree acquainted with the elements of geometry and algebra. Otherwise the lectures cannot impact much instruction though they may possibly afford some amusement (Tylcote 1975: 61).

There is much to be said for the argument, reinforced by many contemporaries, educationalists and historians, that for the first 30 years of the Movement it failed in its aim of attracting the labouring classes. Many institutes closed and then reopened, and this was due to middle-class patronage finding other types of institutions such as philosophical societies that were more acceptable for their own education and interest through theoretical science lectures.

Birmingham Mechanics' Institute, for example, was opened in 1826 and then closed due to the local philosophical society offering lectures in science. In 1825, the newly opened Huddersfield Scientific and Mechanics' Institute in its published rules stated that 'the great object of this Institution is to bring within the reach of all, but more particularly the trading and working classes, the acquisition of useful knowledge – to defuse generally correct principles of Science and Mechanical Philosophy' (*Rules of the Huddersfield Scientific and Mechanic Institute, for the*

Promotion of Useful Knowledge 1825: 1). However, the Institute had a shaky start in its first year, partly as a result of the national banking crisis of 1826, when funds could not be raised to help finance it. It also attracted very few members for whom it was formed, similar to the national trend, and this was probably as a result of having 'scientific and mechanical' as part of the title, giving a strong indication to the public that the objective was to teach specialised science-based subjects.

A Philosophical Society was formed in Huddersfield during the late 1820s specifically for science, and in its first nine months of operation the Committee had organised 24 scientific lectures available to a membership of 310 and it had a library of 1200 volumes (*First Annual Report of the West Riding of Yorkshire Union of Mechanics' Institutes* 1838: 22–3). The society continued for many years, indicating that its membership was made up of mainly educated middle-class professionals with a particular interest in advanced science and mechanics. This reinforces the point that philosophical societies attracted a middle-class educated audience, as opposed to mechanics' institutes which were attempting to attract the industrial working classes 'who were both thirsty for knowledge and anxious to better themselves, and many [of who] saw scientific education as a means of satisfying both these needs' (Peters 1963: 144).

Robert Elliot (1861: 117) observed in 1845 that 'the banquet was prepared for guests who did not come', highlighting the view that the mechanics' institutes were founded for the working classes but were attended by the middling classes. However, this argument is not tenable for the subsequent years, when evidence put forward in Chapter 2 has substantiated that the Movement, having somewhat stalled, continued to expand during the second half of the nineteenth century, the period on which little previous historical research has been undertaken.

Science curriculum and mechanics' institutes

As previously mentioned, scientific lectures, such as those delivered at mechanics' institutes during the 1820s and 1830s, were not new, and it was itinerant lecturers travelling throughout the country that attended mechanics' institutes and philosophical societies (Inkster 1975). Derby Mechanics' Institute, for example, had the support of William Nicol, 'one of the most experienced and skilled of English itinerants in this period' who gave a course of 15 lectures on *The Philosophy of Natural History* to members, one year after the opening of the Institute in 1825. Similar arrangements were made at the Nottingham Mechanics' Institute, founded in 1837, where a well-known itinerant lecturer from London, T. Longstaff, gave 'frequent lecturers on astronomy and meteorology in various English towns' and delivered the first lectures at the Institute which were on chemistry, physiology, silk manufacture, railways, phonology, botany and printing (ibid.: 470).[3] Local dignitaries, particularly the clergy, industrialists or those from the professions, also delivered lectures in their local institutes.

Many of the mechanics' institutes arranged for the local scientific community to support classes and lectures in science. For example, the Mechanics' Institute at Newcastle, which was established in 1824, introduced classes in chemistry, higher mathematics and phrenology. Large audiences were attracted to such public lectures; for example, over 1000 people attended the Manchester Mechanics' Institute to listen to John Davis give a lecture on natural philosophy in 1829 (Inkster 1975: 456).

However, offering scientific lectures and classes did not always attract the kind of members for which mechanics' institutes were established, namely the labouring classes. While most previous studies have concentrated on the larger mechanics' institutes, there were also smaller ones, which were developing in rural as well as urban areas, offering science to a smaller membership. Hitchin Institute in Hertfordshire, for example, which had a population of 5000 in 1851, had a mechanics' institute that was founded in 1835 'for the poor' (Dyer 1950: 114). In 1850, the Institute had 164 members and was ranked 202 out of over 600 institutes in Britain (Hudson 1969). It offered classes in geometry, mechanics, land measurement, hydrostatics, chemistry, geology and botany, and in its first year of operation public lectures were given on woollen manufacturing and the steam engine (Inkster 1975). Other examples of rural institutes which offered science included the Handsworth Woodhouse Mechanics' Institute, established in 1839, and the Bakewell and High Peak Institute in Derbyshire, located in the small market town of that name (Inkster 1975). Offering science subjects, however, did not necessarily encourage large numbers of the working class to attend.

University curricula were still associated with the arts, classics and law, so mechanics' institutes often gained a reputation in their early years for organising public lectures on scientific subjects. However, the depth and relevance of such lectures and classes offered at mechanics' institutes were not stimulating those who perhaps had little or no elementary education to attend, even at relatively cheap rates, as many would have found the subject matter difficult to follow or, indeed, irrelevant to their needs.

This is illustrated effectively by a two-hour lecture given in 1841 at the Sheffield Institute by John Sissons on hydrostatics, which consisted of the following themes:

> Definition of terms, differences between Fluids and Liquids, and pressure of liquids. Hydrostatic paradox explained. Pressure proportional to depth. Consequences for engineering, Water compressible. The application of this principle to Useful Arts. Methods of measuring Solids. Specific Gravities. Definitions. Results (Inkster 1975: 460).

It is thought that the level identified by the themes covered would have been at General Certificate of Secondary Education (GCSE) grades A to C.

A further example was the Darlington Mechanics' Institute Committee, who in 1848 organised several lectures including one on the *Advantages of Knowledge* and a second on *The Apparent Discrepancy, but Real Harmony between the Discoveries of Modern Science*. These lectures were theory-based science-related subjects which required an advanced understanding in order to appreciate them and therefore discouraged the working class from attending, a fact highlighted by Hudson in 1850 (1969) and one of which institute committees themselves were becoming aware as it was having a serious negative impact on membership (*Eleventh Annual Report of the Yorkshire Union of Mechanics' Institutes* 1849: 60).

Hudson was therefore correct, when he observed that the curriculum was 'not suitable or relevant for the mechanic and that it had become the cause of the national movement's troubles by 1850' (1969: p.iv). George Barclay, writing in 1871, made reference to the mechanics' institutes in the period before 1850 and stated that the 'fall of the movement in terms of the incomprehension of members at scientific lectures was due to their lack of elementary education' (Inkster 1976: 279). This was the crux of the problem, for without a good understanding of science, such classes and public lectures were of little interest or relevance to the labouring classes. Shapin and Barnes (1977: 34) state that 'physicians, surgeons and apothecaries; dissenting divines; "enlightened" manufacturers and merchants, having found the cultivation of science appropriate to their own situation in local society, now found compelling arguments for the propriety and value of science for the lower orders'.

Offering such curricula, far from encouraging learning, actually discouraged the 'lower orders' from attending institutes. Therefore, if mechanics' institutes were to survive and support their original aims in supporting working-class education, their committees needed to review the subjects they were offering and at what level, offering elementary education to encourage attendance and support members in progressing to more advanced level subjects in science and mechanics that were practical and not only theoretical.

First, most, if not all, mechanics' institutes throughout the country were offering mechanical and technical drawing from the 1840s, referred to in their reports as drawing, and closely associated with science engineering. Mechanical engineers were indicating that students who had attended drawing classes would benefit from higher wages, as industry required these skills for ongoing success. Manufacturers identified the potential of drawing as a key part of the process of designing new machines and reducing the cost substantially by working out on paper the design and capability of a new machine or product rather than, as previously undertaken, making intricate models (Denis 2001).

Second, taking examples from the Yorkshire Union but also similar nationally, elementary education was being introduced to support adults who had had no formal education. By 1845, Keighley Mechanics' Institute was offering classes in drawing as well as writing, arithmetic, geography and grammar. Writing and arithmetic were the most popular, being taught at elementary level (see Table 3.1).

The Institute also established several advanced classes in science, literature, architecture and mechanical and perspective drawing by 1849 (*Eleventh Annual Report of the Yorkshire Union of Mechanics' Institutes* 1849: 58). The Committee reported that evening classes were 'adequately supplied with teachers and the school-room was well attended by young men anxious to explore the paths of knowledge and to make amends for their neglected education'. These classes were supported by the Department of Science and Arts, who provided certification for those who successfully completed their examinations. The Institute was

Table 3.1 Subjects offered at Keighley Mechanics' Institute and
class size in 1845

Subject	Number in class
Drawing	15
Writing and arithmetic	60
Geography	30
Grammar	15
Total	120

Source: Seventh Annual Report of the Yorkshire Union of Mechanics'
Institutes 1845: 54.

offering a textile class for 20 young women. There were also 28 students attending the French and German classes. In total, there were 345 members attending classes in elementary and advanced subjects (*Eleventh Annual Report of the Yorkshire Union of Mechanics' Institutes* 1849).

In the same year at Keighley, building and construction classes were introduced and attracted students who were apprentices and journeymen from the local building trades. The Committee remarked that it appreciated 'the co-operation of the master builders in the town ... the science and art subjects are taught by authorised masters, assisted by a foreman builder, who gives lessons in practical work'. It was hoped that subjects in other trades could be introduced in the same way (ibid.). By 1851, a specialist drawing class was introduced for men who were training to become masons, joiners or mechanics, supporting the building trades, and the writing, arithmetic and geography classes continued to be popular (*Thirteenth Annual Report of the Yorkshire Union of Mechanics' Institutes* 1851).

A similar pattern of curriculum changes was happening at other mechanics' institutes. Having closed in 1826, the Huddersfield Mechanics' Institute reopened in 1841 and by 1843 was offering classes in pneumatics, natural philosophy, reading, writing, arithmetic, geography, grammar, French, drawing, singing and elocution. Crucially, several of these subjects were offered at elementary level (*Fifth Annual Report of the Yorkshire Union of Mechanics' Institutes* 1843: 25). As a result, there was also a substantial increase in membership, of whom 'nearly the whole of the members are operatives in the receipt of weekly wages' which supports the argument that the Institute, as were others, was offering a relevant curriculum to the working class. The Committee stated in 1844 that 'the attendance is great, the average being 180 to 200 every evening', which is a substantial number of attendees (*Sixth Annual Report of the Yorkshire Union of Mechanics' Institutes* 1844: 27).

The Huddersfield Mechanics' Institute Committee, having failed in 1825, made the strategic decision to concentrate on elementary education with the opportunity for members to progress to higher-level courses knowing that the vast majority had had little or no previous schooling:

> The founders and supporters of this Institution, while providing for the intellectual wants of the adult, have steadily kept in view the importance of

educational training. Their attention has therefore been particularly directed to the efficiency of the classes for elementary instruction (*Seventh Annual Report of the Yorkshire Union of Mechanics' Institutes* 1845: 30).

All classes were 'arranged according to progress' and all members had to first attend the probationary classes before moving up to the more advanced classes (ibid.: 31). The committee made reference to educational training rather than technical education, suggesting that education was the main aim. It would be after 1850 that the term 'technical education' would become more widespread, indicating higher-level courses specific to industry and adult employment.

By offering elementary education and progression onto higher-level subjects, many larger institutes were in a strong position to establish Government Schools of Design, such as those at Bradford, Huddersfield, Keighley and Leeds. The School of Design at Huddersfield, opened in 1846, offered ornamental, architectural and mechanical drawing classes, all of which were 'popular and the standard of work high' (*Eighth Annual Report of the Yorkshire Union of Mechanics' Institutes* 1846: 132).

Frederic Schwann, President, had encouraged the teaching of design and practical chemistry classes at Huddersfield since the Institute's reopening in 1841. He stated that 'the importance of the chemistry class cannot be overlooked in the neighbourhood, when we consider how inferior our fabrics are in beauty of dye and colour, to those of our competitors' (*Fifth Annual Report of the Yorkshire Union of Mechanics' Institutes* 1843: 25). The School of Science offered chemistry classes taught by William Marriott, with the emphasis on practical laboratory sessions supporting the local dye industries. It was described in the *Eighth Annual Report of the Yorkshire Union of Mechanics' Institutes* (1846: 42) as, 'this class, the objects of which are so important ... to so many useful arts in life and its aids being necessary in almost every process of agriculture and manufacture, in the operations of bleaching, dyeing, and printing'. Marriott's successor was George Jarmain, who introduced an advanced course in the chemistry of dyeing, the first in the country (Brown 2005: 2).

Many mechanics' institutes, particularly during the early years when most had to hire accommodation, were not in a position to install laboratories, due to the cost of fitting and lack of space. The Leeds Mechanics' Institute, for example, had a well-established chemistry department by 1847, in response to the needs of the local manufacturers, but did not have laboratories for the practical sessions. Until the Institute had funds to equip its own laboratories later in the year, students were expected to carry out experiments in their homes. An advanced class of chemical manufacturing was introduced and supervised by a teacher in the newly opened laboratory. 'The practical method will qualify students in the short term for analytical research, develop talent where it exists, and render the services of each pupil more valuable to their employers and profitable to themselves'. This indicates that on completing elementary education, students were confident, as well as able, to go on to be successful in the advanced classes (*Ninth Annual Report of the Yorkshire Union of Mechanics'*

Institutes 1847: 59). Similar developments were taking place at institutes across the country.

The School of Design at Keighley was offering classes in the practical study of design and drawing by 1849, for 'those students who are engaged in the fancy manufactures of the district and to those who are connected with ornamental trades and mechanics'. The mechanical drawing class was delivered through the School and students 'acquired considerable skill in drawing, and a good knowledge of the mechanical operations of steam, and the movements of locomotives and other engines'. The Committee highlighted that these classes helped to 'greatly improve their arithmetical and commensurate knowledge' (*Eleventh Annual Report of the Yorkshire Union of Mechanics' Institutes* 1849: 57–8).

The Institute also identified the importance of developing the skills of its members though the classes and curriculum being offered:

> in a manufacturing community like this, where much ingenuity is required to invent new patterns, it is exceedingly desirable to elicit and cultivate native talent, in order that successful competition may be carried on with surrounding towns in the worsted manufactories (ibid.: 63).

This observation by the Committee was being made some two years before the Great Exhibition of 1851, a crucial watershed in adult education and training, and it highlights how institutes had already begun to recognise the importance of relevant curricula being offered to the working-class memberships, employed by manufacturers who benefited from their workers' education.

It was not just mechanics' institutes in the textile districts that were offering relevant curricula. At Filey, for example, in 1849,

> 'despite having few members of the Aristocracy to support the mechanics' institute [financially], the town has acquired some accommodation and is offering courses to fishermen and their families who spend the evenings between December and January reading and writing; the only time of the year they can afford the time to do so.

There were up to 20 fishermen, of whom several were around 60 years old and who attended the evening winter classes as fishing was impractical due to regular bad weather and lack of accurate navigation equipment (*Eleventh Annual Report of the Yorkshire Union of Mechanics' Institutes* 1849: 57–8).

Conclusion

By 1850, using Huddersfield as a case study, the profile of those who attended the mechanical and architectural drawing classes at the Mechanics' Institute included 'mechanics, operatives, workers in brass and iron, builders, joiners, cabinet makers, carvers, gilders and painters'. Students of ornamental drawing had become competent in cloth designs and colour, no doubt relying on the knowledge and developments in the dye industry supported by the Institute's chemical classes (*Twelfth Annual Report of the Yorkshire Union of Mechanics' Institutes* 1850: 40). The Huddersfield Committee, like those at institutes in other parts of the country, continued to make specific mention of the fact that chemistry was of 'great importance to manufacturers and to the arts of bleaching and dyeing', rendering the class 'a real and solid acquisition to the Institution' (*Thirteenth Annual Report of the Yorkshire Union of Mechanics' Institutes* 1851: 52). The lessons were of a practical nature and students were expected to carry out their own experiments. Huddersfield was not the only institute offering courses that specifically supported the textile industry. Almost all towns and settlements that had a mechanics' institute in the North offered these subjects, which supported the needs of their local industry. As a result, many mechanics' institutes had weathered the storm of decline by the early 1850s, albeit with low numbers, closures and then reopening following responsiveness to the needs of members with subjects that were relevant to their requirements and those industries they worked in, underpinned with elementary education.

Notes

1 Wilson had lectured at the Edinburgh School of Arts and Phillips was a popular lecturer at the London Mechanics' Institution.
2 Dalton came to Manchester in 1793 to teach mathematics at the dissenting Manchester Academy and was elected to the Literary and Philosophical Society; later he became an active committee member of the Manchester Mechanics' Institution.
3 In the case of the Yorkshire Union, as in other parts of the country, itinerant lecturers supported several institutes, including the one at Barnsley where the Institute offered public lectures and classes on chemistry over several years after its formation. At the Ackworth Mechanics' Institute, Charles Morton of Sheffield delivered several lectures on the application of science to agriculture and manufacturing, and at Huddersfield, Rose of Edinburgh gave a series of lectures on geology and mineralogy (Inkster 1975: 456–7).

References

Barnard, H. C. (1966) *A History of English Education from 1760* (London: University of London).

Brown, R. (2005) 'Colourful chemistry in a northern town', *Chemistry World*, unpublished paper (London: Royal Society of Chemistry): 2.

Cotgrove, S. F. (1958) *Technical Education and Social Change* (London: Allen and Unwin).

Denis, R. C. (2001) 'An industrial vision: the promotion of technical drawing in mid-Victorian Britain', Purbrick, L. (ed.) *The Great Exhibition of 1851, New Interdisciplinary Essays* (Manchester: Manchester University): 53–78.

Dyer, L. J. (1950) 'The Hitchin Mechanics' Institute', *Adult Education* Vol. XXIII, No. 1, 113–121.

Elliot, R. (1861) 'On the working men's reading rooms, as established in 1848 at Carlisle', *Transactions of the National Association for the Promotion of Social Science*, 110–117.

Griscom, J. (1824) *A Year in Europe, Comprising a Journal of Observations in 1818 and 1819* (Second edn. New York: Collins).

Haigh, E. A. H. (ed.) (1992) *Huddersfield: A Most Handsome Town* (Huddersfield: Kirklees Library Service).

Huddersfield Scientific and Mechanic Institute (1825) *Rules of the Huddersfield Scientific and Mechanic Institute, for the Promotion of Useful Knowledge* (Huddersfield: T. Kemp, University of Huddersfield Archives).

Hudson, J. W. (1851) *The History of Adult Education in which is comprised a Full and Complete History of the Mechanics' and Literacy Institutions* (Reprint. London: Woburn 1969).

Inkster, I. (1975) 'Science and the Mechanics' Institutes, 1820–1850: The case of Sheffield', *Annals of Science* No. 32, 451–474.

Inkster, I. (1976) 'The social context of an educational movement: A revisionist approach to the English Mechanics' Institutes, 1820–1850', *Oxford Review of Education* Vol. 2, No. 3, 277–307.

Inkster, I. (1980) 'The public lecture as an instrument of science education for adults – the case of Great Britain, c.1750–1850', *Paedagogica Historica: International Journal of the History of Education* Vol. 20, No. 1, 80–107.

Peters, A. J. (1963) 'The changing idea of technical education', *British Journal of Educational Studies* Vol. 11, No. 2, 142–166.

Salt, J. (1966) 'The creation of the Sheffield Mechanics' Institute', *The Vocational Aspect of Education* Vol. 18, No. 40, 143–150.

Shapin, S., and Barnes, B. (1977) 'Science, nature and control: Interpreting Mechanics' Institutes', in *Social Studies of Science* Vol. 7, 31–74.

Tylecote, M. (1974) 'The Manchester Mechanics' Institution, 1824–1850', Cardwell, D. S. L. (ed.) *Artisan to Graduate, Essays to Commemorate the Foundation in 1824 of the Manchester Mechanics' Institution, Now in 1974 the University of Manchester* (Manchester: Manchester University Press): 60–61.

West Riding of Yorkshire Union of Mechanics' Institutes (1838) *First Annual Report.*

Yorkshire Union of Mechanics' Institutes (1843) *Fifth Annual Report.*

Yorkshire Union of Mechanics' Institutes (1844) *Sixth Annual Report.*

Yorkshire Union of Mechanics' Institutes (1845) *Seventh Annual Report.*

Yorkshire Union of Mechanics' Institutes (1846) *Eighth Annual Report.*

Yorkshire Union of Mechanics' Institutes (1847) *Ninth Annual Report.*

Yorkshire Union of Mechanics' Institutes (1849) *Eleventh Annual Report.*

Yorkshire Union of Mechanics' Institutes (1850) *Twelfth Annual Report.*

Yorkshire Union of Mechanics' Institutes (1851) *Thirteenth Annual Report.*

4 Scientific and technical education 1851–1900

Introduction

Fieldhouse (1998) highlights that for most of the nineteenth century until the passing of the Technical Instruction Act in 1889 there was little government support for technical education. He refers to the belief that the early success of industrialisation without formal, and technical, education meant that there was a general complacency regarding government support. It was, in any case, the period of liberal *laissez-faire* philosophy that discouraged any state intervention. In contrast, European industrialisation occurred under the protection of the state with what Green (1995: 128) points out was supported with technical and scientific education 'from the centre as an essential adjunct of economic growth'; important for those countries wishing to challenge Britain's industrial lead.

Following the repeal of the Statute of Artificers in 1814, whilst there were apprenticeship schemes, they were practical, with only 'on the job' training. They were therefore easy to organise but inadequate, as having no grounding in education or practical skills. However, as discussed in Chapter 3 of this volume, the mechanics' institute movement had identified by the 1840s the need to offer both elementary education and work-related subjects such as those offered to the textile industry. By the late 1840s and early 1850s, technical education was superseding 'one-off' high-level scientific public lectures.

The transformation from scientific to technical education was not straightforward. Roderick and Stephens (1970: 103) believed that there were specific problems with offering technical education during the second half of the nineteenth century. They point out that technical education did not have the same status as classical education being offered in the ancient universities. It was aimed at 'providing instruction in the sciences underlying the arts', that is, aimed at developing the manual and technical skills required to work with tools and machinery. They argue that the subject content was not advanced enough to support technological developments in support of Britain's rapidly developing industrialisation, and not of a practical nature to support skilled or semi-skilled workers in their workplace. Education offered before the late 1850s in the mechanics' institutes was more commonly known as 'useful knowledge' (Musgrave 1964).

Pre–1851 exhibitions

One important way of publicising technical developments that was developing was through exhibitions held at many mechanics' institutes. The idea for an international exhibition to be held in London came from the many successful events that had been held in Britain from the late 1830s at various institutes. These exhibitions were seen as 'enlightening the public and awakening their curiosity' and at the same time provided publicity and raised much-needed funds for the institutes (Tylecote 1957: 78). Their committees were confident 'that exhibitions would attract the working classes, stimulating their imagination and making them aware of new arts and science developments'.

The first large exhibition was held at the Lancashire and Cheshire Union Manchester Mechanics' Institute in 1837. Exhibits included 31 model steam engines, 79 models of 'useful machines and ingenious mechanical contrivances', 12 models of public buildings, 90 philosophical [scientific] instruments, 140 India ink and coloured designs and drawings, 28 specimens of painted and stained glass and 10,000 insects (ibid.: 306).

In the case of the Yorkshire Union, the exhibition held at the Bradford Mechanics' Institute in 1839 raised between £700 and £800, the proceeds going towards a new Institute building (ibid.: 229). In the same year at Halifax, an exhibition on science and art was jointly organised between the Infirmary, the Literary and Philosophical Society and the Mechanics' Institute, which attracted 100,000 visitors (ibid.: 238). Other Yorkshire Union mechanics' institutes that organised exhibitions included Todmorden in 1839 (*First Annual Report of the West Riding Union of Mechanics' Institutes* 1839: 24). At Sowerby Bridge, an exhibition held in the same year lasted seven weeks, attracted 29,000 visitors and made a profit of £142 (ibid.: 38). An exhibition of arts and manufactures held at the Leeds Institute in 1842 raised £1630, which supported the purchase of a building and helped pay off some of the debt inherited from the previous Literary Institution that it had taken over (ibid.: 71). At the Huddersfield Institute, an exhibition was held in 1844 and attracted between 500 and 600 visitors. It included displays of 'microscopes, dissolving views and optical illusions' and a series of experiments including several on the science of galvanism as well as 'demonstrations on the use of oxy-hydrogen blow pipes, air pumps, a diving bell and working models of machinery' (*Sixth Annual Report of the Yorkshire Union* 1844: 27).

Elliot (2004: 398) identified that 'one of the most important manifestations of public science in Derby was the exhibitions organised at the Mechanics' Institute' which, along with the local museum, 'attracted tens of thousands of visitors from the Midland region and further afield'. The Birmingham Mechanics' Institute opened its exhibition to foreign competitors in 1849. It highlighted the serious concerns that Britain was lagging behind other countries in relation to industrialisation and skill. In part, at least, this was due to the lack of technical education available to the majority of employees (Davis 1999: 11).

The 1851 Great Exhibition of the works of industry of all nations

Dunstan (1996: 11–12) believes that 'The Great Exhibition therefore had its antecedents in the modest provincial exhibitions organised by mechanics' institutes' over the previous twenty years or so. Such 'exhibitions were a straightforward application of learning by looking' during a period when some working people were just beginning to receive an elementary education.

The Great Exhibition was opened in London by Queen Victoria on 1 May 1851. It was the idea of the Queen's consort, Prince Albert, who, with other organisers, saw such an Exhibition as a means of improving design and artisanal skills in support of Britain's industrialisation in response to foreign advancement and competition in science and technology. As President of the Society for the Encouragement of Arts, Manufactures and Commerce (hereafter referred to as the Society of Arts), Prince Albert had substantial backing for the idea and support for a Great Exhibition (Walker 2015). Lord Henry Brougham, at the time the Whig Chancellor, supported educational reforms in Parliament. He wrote articles on scientific education in the influential Edinburgh Review, where he once stated that 'British artisans were the least trained and the middle-class manufacturers the worst educated in Europe' (Auerbach 1999: 11).

Prince Albert believed that the involvement of the working classes 'was critical to the success of the exhibition'. He included working men on the Exhibition Committee 'to promote the interests of the working classes'. Samuel Wilberforce, the Bishop of Oxford, who was on the London Committee of the Exhibition, gave a speech in which he stated that 'it [the Great Exhibition] sets forth in its true light the dignity of the working classes ... it tends to make other people feel the dignity which attaches to the producers of these things [exhibits]' (Auerbach 1999: 129).

In the year prior to its opening, industrial towns throughout Britain supported the Great Exhibition through local committees which were established to advertise the event and support financially those employees wishing to attend, assuming their employers supported this. In the case of Manchester, for example, the Unitarian industrial families of Heywood, Philips, Henry, Potter and Greg, who were on the town's committee, also supported adult working-class education. They had all been influential in the establishment of the Manchester Mechanics' Institute as well as their own local institutions (Auerbach 1999).

Barton (2000: 152) has identified that the Manchester Committee included working men, with two men from each principal workshop and factory to assist in supporting the Exhibition. The purpose of such committees, established all over the country, was:

> to originate an active canvass amongst the artisans in our different machine shops and manufactories to ascertain how many individuals, or associated bodies, will prepare specimens of their skill for the exhibition. A further object of the committee was to arrange for a cheap trip on a series of days, so as to allow all interested to visit the exhibition at the lowest possible cost.

The Southampton Committee listed all donations in support of the Great Exhibition and among them were clerks, draftsmen, workmen and servants, which illustrates the cooperation and support that came from the working classes. At Bolton, the Committee created a separate subscription, called the 'operative fund', to assist artisans in 'perfecting items for display at the exhibition' (Hobhouse 2002: 135). Leeds mill owners were encouraged to support its Committee by encouraging their workers to subscribe. John Gott's employees donated £75 towards the Great Exhibition (ibid.). At Huddersfield, a donation of £25 was given by the largest local landowner, Sir John Ramsden, that enabled eight students from the Mechanics' Institute to attend the Exhibition (*Thirteenth Annual Report of the Yorkshire Union* 1851: 60). It is no coincidence that the Great Exhibition of 1851 was a contributing factor to the Mechanics' Institute Movement. The establishment of local committees for the purpose of fundraising were strongly associated with mechanics' institutes.

Over six million people visited the Great Exhibition between May and October 1851 and as many as 110,000 visitors attended on October 7 alone, presumably wanting to see the exhibits before the Crystal Palace closed at the end of that week. It took 'a second hackney carriage to take the day's takings [one shilling entrance fee] to the Bank of England' (Hobhouse 2002: 69).

Barton (2000: 150) makes reference to the fact that mechanics' institutes were inspirational for the Great Exhibition, because exhibitions on a smaller scale had been 'held in several provincial towns between 1838 and 1840'. She also believes that the Great Exhibition would put to an end:

> the contempt shown for tradesmen and mechanics once the world could witness the skill involved in the production of artefacts for display. Many believed that the event would provide the ideal opportunity for working men to demonstrate not just their skills and intellectual capabilities in the design and making of exhibits, but also their respectability.

Gurney (2001: 116) believes that the relationship between the working class and the Great Exhibition was widely discussed at the time and that the liberal intellectuals hoped that it 'would fulfil a wider educative function and exert a civilising influence on the majority'. Gurney highlights that 'manual workers have now achieved a recognition and respect in society': the Exhibition was in fact the first

> public national expression ever made in this country ... a marvellous display of the trophies and triumphs of labour [which] could not fail to fill working men with pride and inspire them with a sense of their position in the State (ibid.: 117).

The exhibits themselves provided insight into technological developments and engineering knowledge. They also encouraged mechanics' institute committees to consider including similar examples in class instruction. The exhibits were put into one of four categories, as shown in Table 4.1.

Table 4.1 Classification of a selection of exhibits at the Great Exhibition

Category	Exhibits
Raw materials	Mining and Quarrying, Metallurgy and Mineral Products, Chemical and Pharmaceutical Processes and Products.
Machinery	Machines for Direct Use (Carriages, Railways), Manufacturing Machines and Tools, Mechanical Engineering, Architectural and Building, Naval Architecture, Military Engineering, Ordnance Armour, Agricultural and Horticultural Machines and Implements, Philosophical Instruments [scientific] and Miscellaneous.
Manufactures	Cotton, Woollen and Worsted, Silk and Velvet, Manufactures from Flax and Hemp, Woven, Spun, Felted and Laid Fabrics, Mixed Fabrics, including shawls, Leather, Skins, Fur and Hair Paper, Printing, Bookbinding, Cutlery, Edge and Hand Tools, and Surgical Instruments, Glass, Ceramic Manufacture, Decorative Paper, Paper Hangings, Miscellaneous Manufactures and Small Ware.
Fine art	Fine Arts, Sculpture, Mosaics and Enamels.

Source: Auerbach 1999: 93.

By 1850, a year before the Great Exhibition, there were over 600 mechanics' institutes in England and Wales alone with a membership of nearly 85,000. Having already highlighted that, after 1850, there was growth rather than decline, research indicates that the Great Exhibition was by far the single most important event that turned adult education from being philosophical and elitist for the educated, to scientific and technical, with elementary education underpinning it, for the masses. It encouraged and enthused employers and employees to support it through making exhibits and visiting the Exhibition, which encouraged further technical and scientific interest and development in British manufacturing over the following decades.

The success of the Exhibition encouraged many mechanics' institutes throughout the country, such as Leeds, Liverpool, Halifax, Huddersfield and Nottingham, to hold annual exhibitions of manufactured goods, which, apart from raising additional income for their institutes, also raised the profile of new inventions and patents, some of which were made by members themselves, such as a wool dust extractor invented by Thomas Broadbent, a former student in 1857 at the Huddersfield Institute.

Barton (2000: 162) believes that the Great Exhibition reassured the country that the working classes were no longer a threat to industrialisation, as their 'involvement in all aspects of the Exhibition showed that the risk of insurrection was over and an era of incorporation, collaboration and reformist politics had commenced'. This was therefore a supportive climate in which mechanics' institutes would be able to flourish. Two years after the closure of the Great Exhibition, its impact on working-class technical education was still being felt. The Government established the Department of Science and Art in 1853, which offered qualifications, prizes and funding for science and art classes where quality teaching was being offered for mechanics' institutes and similar. The Great Exhibition invigorated the Mechanics' Institute Movement. It brought to the fore the urgent need for scientific and technical education in support of Britain's continued industrialisation, and mechanics' institutes reacted accordingly. As Harrison (1961: 213) noted, institutes were saved from bankruptcy 'by two timely developments – the examination system of the Society of Arts, and the demand for technical and scientific education'.

James Hole (1853: 44), influenced by the findings of the Exhibition, had particular interest in reforming adult education amongst the working classes. Writing in 1853, he stated that 'education is not an affair of childhood and youth; it is the business of the whole of life'. He went on to say that 'the nation which possesses the largest number of skilled artisans, capable of availing themselves of the aids which science lends to industry, will, other things being equal, be the richest nation' (ibid.: 47). Hole had identified the importance in both rural and industrial areas of mechanics' institutes in supporting adult working-class education. He believed that the rural institutes could provide courses in the science of agriculture for farmers and husbandmen, supporting 'the culture of land, the maturing of crops, their value when reaped, the feeding and treatment of stock, the manufacture and management of butter and cheese' (ibid.: 51). Hole also saw the importance of chemistry as an industrial subject supporting the dyeing, bleaching and other trades in support of British industrial progress.

Hole also identified the need for institutes to have qualified teachers, newspapers and reading rooms, social gatherings, exhibitions, penny savings banks and itinerating libraries. The government, he believed, should take responsibility for funding and making available the Society of Arts examinations and certificates, in recognition of working-class adults' educational achievements. Of all his requests, the only one not to develop after 1850 was a national Union of Mechanics' Institutes, although the Society of Arts did support mechanics' institutes throughout the country, which together formed the Society of Arts Union for its administration of examinations. Hole's publication and lecture tours, along with others, inspired the second period of growth in the movement after 1850 with the potential to support adult education for the masses, which was both relevant and supportive of rural and industrial developments.

Following the findings of the Exhibition, the Society of Arts debated the concern that the development of science and technology in Britain was being overtaken by other countries. Members of the Society, which included Whigs, Tories, Radical civil servants, aristocrats, industrialists, manufacturers and academics, despite their diverse political views, agreed that there should be a national system of compulsory education and 'adult remedial courses for those who already lacked schooling'. Without this foundation, the workforce would have little understanding or knowledge of 'scientific elements to their trades' (Barton 2000: 144).

The Society also stated that in order to support these needs, technical and teacher training schools should be established to teach new, specialised skills necessary to operate modern industrial machinery and develop a scientific knowledge in relation to science and industry. The Society also believed scientific and technological developments could be further supported through the building of additional government Schools of Design, first established in 1832. The Society also noted:

> That the success of literary and scientific institutions and mechanics' institutes, in the cultivation of literature, science and fine arts, and in the diffusion of useful knowledge, might be powerfully promoted by the combination of many institutions in a Union with the Society of Arts on the basis of perfect security to the continued independence of the institutions (Garner 1985: 257).

By 1856 there were 340 literary and scientific institutions and mechanics' institutes in the Union. Institutions had to be members in order to offer the Society's examinations (*Society of Arts Manuscript Subscription Books* 1852–1856). The Society of Arts examinations were held in London, and at Huddersfield in 1857 for the first time as the northern centre for conducting Society of Arts examinations. It was selected for its reputation in administrating other external examinations and good teaching (*Huddersfield Minutes of the General Committee* 25 November 1857: 7; *Nineteenth Annual Report of the Yorkshire Union of Mechanics' Institutes* 1857: 84; Walker 2008). Previously, it had only been possible to sit their examinations in London. The following year and thereafter, Society of Arts examination centres were established across all parts of the country.

National examinations were first introduced through the government's Science and Art Department, South Kensington, which was established in 1853. It was administered by the Board of Trade, which offered examinations in basic science and industrial art. The Science and Art Department and the Society of Arts both offered external certification in a variety of technical subjects by the mid-1850s. In the case of the Society of Arts, a Union of Institutions was formed and local boards administered the examinations on behalf of the Society of Arts. Offering examinations, with government grants, further supported institutes both financially and academically.

The Lancashire and Cheshire and the Yorkshire Unions of Mechanics' Institutes offered examinations until the twentieth century, and the Northern Union had been interested in offering its own examinations. This was partly due to the cost of sitting examinations in London at the Society of Arts headquarters. There was also a sense of regional influences with regard to 'a high standard of knowledge in the persons by whom the examinations are conducted ... without disrespect to the metropolitan examiners' (Northern Counties Technical Examinations Council [NCTEC] 1970: 11–12). In the case of the Northern Union, there was a proposal put forward in 1851 to offer examinations, and by 1856 subjects offered, syllabuses and appointments of examiners were all in place. It seems probable that with the establishment of the Department of Science and Art as well as Society of Arts examination schemes, the Northern Union was unable to compete (NCTEC: 11–13). However, the Union had provided a foundation on which the Northern Counties Technical Examinations Council

(NCTEC) was established in 1920 and then became the Northern Council for Further Education (NCFE) in 1981, and which continues to offer vocational qualifications. The Union of Lancashire and Cheshire Institutes, on the other hand, offered examinations in technical subjects well into the twentieth century. It is probably no coincidence that Percival Sharp, former President of the Union, became Director of Education for Newcastle between 1914 and 1919 and instigated the establishment of the NCTEC (Northern Counties Technical Examinations Council 1945).

The Lancashire and Cheshire Union institutes were also offering examinations. At the Manchester Institute by the 1860s, for example, elementary subjects were being examined by the Union. At the same Institute, other subjects, such as bookkeeping, were examined by the Society of Arts. As elsewhere, students sitting Science and Art Department examinations brought in extra revenue to the Institute if they passed them. Equipment and scientific apparatus was provided by the Department to institutes that offered its examinations (Cruickshank 1974: 141). The Manchester Institute also 'seized the opportunity afforded by the new City and Guilds of London Institute to start in 1880 classes in bleaching, dyeing and printing, cotton manufacture, mechanical engineering, iron and steel, fuel, telegraphy and carriage building' (ibid.: 145).

Wright (2001) identifies that by the late 1860s, institutes that were offering the Department of Science and Art courses were provided with grants for teachers where quality classes in the sciences and arts. The Society of Arts[1] and the Department of Science and Art examinations brought stability to the mechanics' institutes. William Gladstone, then the Liberal Chancellor of the Exchequer and later Prime Minister, addressing a meeting of the Lancashire and Cheshire Institutes in 1862, spoke of:

> the age of examinations ... it is not too much to say that the experience of half a century, as well in the Universities ... appears to have shown that the method of examination is best, and perhaps the only method, by which in England of the nineteenth century any due efficiency can be imparted to the general business of education (*Manchester Guardian* 8 February 1862).

This was certainly the case at Keighley, which, like others, was to some extent ahead of the game. In 1849 the Committee at Keighley Institute had identified the importance of developing skills though its classes.

> In a manufacturing community like this, where much ingenuity is required to invent new patterns, it is exceedingly desirable to elicit and cultivate native talent, in order that successful competition may be carried on with surrounding towns in the worsted manufactories (*Eleventh Annual Report of the Yorkshire Union* 1849: 63).

By the 1860s, the Institute, which included a School of Art, had developed and expanded substantially from its original establishment in 1825 by four unskilled labourers, a reed-maker, a painter, a tailor and a joiner (Tylecote 1957). It was recognised by the Science and Art Department in Whitehall as having a national reputation not only for art but also science, and in consequence the large number of medals and prizes awarded to its students. As a result, government funding was made available and the Institute received a grant of about £500 a year. The annual report of 1866 noted that 'the Keighley School is not yet surpassed, if equalled by any school in England' (*Twenty-Eighth Annual Report of the Yorkshire Union* 1866: 58). The report also made reference to the evening science classes at Keighley which were generally satisfactory, with students attending the science and art classes during the day doing substantially better. It may have been the practical nature of chemistry in relation to dyes and associated chemicals used in textiles that made the subject relevant and interesting to those attending during the day. However, those attending the evening classes often did so after a full day's work (ibid.: 59). By 1880, there were over 70 mechanics' institutes offering examinations through the Department of Science and Art to about 7000 students, of whom 4000 were taking science subjects and the remaining 3000 attending art and design classes (Stephens and Roderick 1972: 357).

The City and Guilds of London Institute for the Advancement of Technical Education, hereafter referred to as City and Guilds London Institute, was established in 1878 and was supported by 16 London City Livery Companies to establish a national scheme for technical education (Curtis 1967: 495). Examinations in technical subjects were offered from 1881 with just 116 candidates sitting at 110 examination centres, of which at least 45 were mechanics' institutes or similar institutions. The following year, the number of centres had increased to 134, of which 62 were mechanics' institutes or associated institutions (City and Guilds 1993: 27–8). During the twentieth century, City and Guilds London Institute became responsible for all examinations in technical subjects, leaving the Society of Arts to concentrate on offering examinations in commercial subjects (Argles 1964: 148).

Where there was a demand for skilled and educated workmen, the institutes responded by offering technical education with public-recognised examinations through City and Guilds London Institute and the Society of Arts. Predominately, these activities took place in the North of England (Royle 1985: 198). In the South, particularly in the more rural areas, there was less interest in technical education and the institutes tended to be libraries and scientific societies, with an emphasis on the loan of books, general lectures and social activities, rather than on formal classes and examinations, a point which has been assumed by many educationalists and historians to be nationwide.

The Keighley Weaving School had entered four students for the City and Guilds Examinations, two gaining certificates (*Forty-Fourth Annual Report of the Yorkshire Union of Mechanics' Institutes* 1882: 111). The Cotton Manufacture Course at Hebden Bridge Institute in 1882 was delivered through City and Guilds and was taught by a former student of the science class. The Committee believed passionately in students taking up the opportunity to sit examinations who were 'in the trades and manufacture of the district ... to take advantage of the City and Guilds examinations, for the technological certificates which will doubtless become valuable testimonials to them in seeking employment' (*Forty-Fifth Report of the Yorkshire Union of Mechanics' Institutes* 1882: 110). Thus, the Mechanics' Institute Movement was providing opportunities for those sitting examinations to have national recognition for their studies, and supporting their employment prospects.

The 1870 Education Act

The Great Exhibition of 1851 had highlighted serious competition from overseas. Developing technical education was paramount in supporting this. However, British exhibits at the Paris Exhibition in 1867, only some 16 years later, continued to be sub-standard in comparison to those from other countries. At the 1851 Exhibition there were 100 different categories in which almost all British manufacturers were superior, whereas at the 1867 Exhibition Britain was only ahead of her competitors in ten of them (Curtis 1967: 495). While the Department of Science and Art and the Society of Arts were firmly established, there was still some way to go in developing technical education, underpinned with elementary education in the three 'Rs'.

William E. Forster, after whom the 1870 Education Act was named, was Liberal MP for Bradford and President of the Bradford Mechanics' Institute. He also sat on several institute Committees in the West Riding of Yorkshire, including the one at Huddersfield Institute. Forster established his own woollen industrial community in Burley-in-Wharfdale in Yorkshire, which bore a close similarity to a similar model village founded by Sir Titus Salt at Saltaire near Bradford. Burley had its own mechanics' institute. Forster was also involved in the foundation of the London Working Men's College as well as being closely involved with the Halifax Working Men's College, the first one to be built outside London (Baker 2001). Forster identified that industrial prosperity depended on 'the speedy provision of elementary education' (Baker 2001: 225). Under the 1870 Act elementary education was to be offered through permitting local rates to be raised in support of their local School Board. The Act gave powers to the Boards to make education compulsory in their area if they so wished (Curtis 1967: 279).

Many mechanics' institutes had been delivering elementary education to adults since the 1840s and 1850s and in many cases to children, too. Indeed, the Dean of Hereford, Richard Dawes, had given the annual address at Huddersfield in 1855 and identified that, as schools were few and far between, mechanics' institutes should introduce elementary education (Dawes 1855). This arrangement must have influenced Foster's thoughts, as he himself was involved in the movement. Some institutes, amongst them Manchester, felt that the introduction of local school boards under the Act, giving them responsibilities for elementary education, meant that such institutions could concentrate on adult elementary and advanced education. Where towns did not have an arrangement for school boards for several years, mechanics' institutes continued to offer elementary education to children, such as those at Huddersfield and nearby Slaithwaite in the

West Riding of Yorkshire (Twenty-Second *Annual Report of the Yorkshire Union* 1860).

Royal Commissions

In response to the ongoing challenges relating to adult education, a government Select Committee on Scientific Instruction produced one of several reports which contributed to the Devonshire *Report of the Royal Commission on Scientific Instruction and the Advancement of Science* which was published in 1875. The Chair, William Cavendish, the seventh Duke of Devonshire, was also patron of the Yorkshire Union of Mechanics' Institutes and supported several institutes in Yorkshire and Derbyshire. The report took the form of a detailed survey of scientific education at universities and other institutions. It urged that children in the elementary schools should have more science teaching and training colleges should provide new courses for science teachers. The Education Department and the Science and Art Department should be coordinated and work more closely together. The report also made recommendations for the training, recognition and payment of qualified science masters, and for building grants for certain kinds of institutions (Maclure 1969: 106). The findings of the Royal Commission resulted in the passing of the 1889 and 1890 Technical Instruction Acts.

The ninth Duke of Devonshire had a particular interest in science and was a critic of 'the undemanding educational regime practiced in most public schools'. He too, as an MP, was involved with government education policy that resulted in the passing of the Balfour 1902 Education Act which, amongst other things, resulted in a rapid growth of secondary schools (Parry 2015).

Further select committee reports resulted in the *Report of the Royal Commission on Technical Instruction* (the Samuelson Report), which was published in 1884. Bernard Samuelson, the Chair, had been an ironmaster and engineer prior to becoming an MP in 1859 and had a personal interest in technical instruction, travelling throughout Europe making comparisons. Sir Swire Smith, who was one-time President of the Keighley Mechanics' Institute, was also on the Committee. Smith, a woollen manufacturer in the town, published a paper in support of technical education after visiting Europe with a number of colleagues. His findings contributed to the Royal Commission on which he sat (Smith 1886: 16). Both men were aware of foreign competition and wanted technical education to support Britain's economic position in the industrial world (Maclure 1969: 121). The Commission was set up to investigate the training given in technical institutions and science teaching from elementary to advanced level. The report emphasised the importance of local authorities providing first-class technical instruction in a variety of educational establishments, including day schools and mechanics' institutes, to improve quality and developments in manufacturing. Betts (1998: 272) refers to the Commissioners having to pay their own expenses, suggesting that the government had little interest in the report. Nevertheless, Ward (1973: 34) believes that the Report of the Royal Commission was a watershed in support for technical education.

Both Commissions identified the lack of commercial training offered in mechanics' institutes. Swire Smith had brought evidence that France and Germany were more advanced than Britain in vocational training. Various mechanics' institutes across Lancashire and Yorkshire, as elsewhere, were offering vocational commercial classes, especial in phonology, including at Huddersfield as early as 1846, Lockwood, near Huddersfield, Bacup, Bolton, Burnley and Wigan, all offering shorthand and bookkeeping. The Manchester Institute had thought about setting up a separate college of commerce in 1882 and Sir Isaac Pitman delivered classes, using his phonetic shorthand, between the 1850s and 1877 in the town (Hemming 1978: 41–2).

The Union of Lancashire and Cheshire Institutes and the Yorkshire Union of Mechanics' Institutes offered examinations in commercial subjects until the twentieth century. In the case of the former, by 1891 it was offering commercial geography and a new subject, typewriting. In the annual report of that year the course was described as 'rapidly taking its place as a most popular subject, particularly among females' (*Fifty-Third Annual Report of the Union of Lancashire and Cheshire Institutes* 1891: 165). The Yorkshire Union also offered its own examinations; students were able to gain design and weaving qualifications that were recognised by the Worshipful Company of Clothworkers. The Company donated scholarships enabling several students from across the Union to continue their studies at the Yorkshire College, Leeds, later to become Leeds University (*Forty-Second Annual Report of the Yorkshire Union of Mechanics' Institutes* 1880: 104).

Technical Instruction Act and Local Tax Act

The result of these Royal Commissions was the passing of the Technical Instruction Act of 1889 which gave local authorities the power to levy a penny rate in order to provide technical courses, appoint teachers and provide grants to technical schools and mechanics' institutes. Marsden (1969: 44) states that the Act was the first one that was a useful piece of parliamentary legislation passed in Britain in support technical education. The Act specified:

> Instruction in the principles of science and art applicable to industries, and in the application of specific branches of science and art to specific industries or employments. It shall not include teaching the practice of any trade or industry or employment, but, save as aforesaid, it shall include instruction in the branches of science and art with respect to which grants for the time being made by the Department of Science and Art, and any other form of instruction (including modern languages and commercial and agricultural subjects) which may for the time being be sanctioned by the Department by a minute laid before Parliament and made on the representation of a local authority that such a form of instruction is required by the circumstances of its district … manual instruction shall mean instruction in the use of tools, processes of agriculture, and modelling in clay, wood, or any other material (Technical Instruction Act 1889, Section 8, in Musgrave 1964: 107–8).

The key point was that the Act was to aid 'instruction in the principles of science and art applicable to Industry and agriculture (Musgrave 1964: 179).

In 1890, the government, through the passing of the Local Tax Act, put a tax on wines and spirits, and it was decided that the money raised should be used for supporting technical education. This became known as 'Whisky Money', which did provide some funding for mechanics' institutes and other institutions. Fieldhouse (1998: 43) identifies that with the passing of the 1889 Technical Instruction Act and the Local Taxation Act of 1890, both of which raised government revenue for education, £750,000 was raised by 1891, which was increased to over £1 milllion by 1900, providing state-funded adult education. When the tax ceased in 1902, technical education was already well established through central funding from government (Curtis 1967: 497).

While developments in technical education post–1850 were in many ways quite substantial, they were not perfect. Roderick and Stephens (1970: 103) point out that instead of producing an 'officer class' of researchers and managers in support of industry, and therefore raising the status of technical education, it was 'aimed at the commercial classes and at skilled and unskilled workers'. They argue that the level of qualifications were no higher than secondary school level until 1900. They also argue that technical education was offered through evening classes well into the twentieth century and that there were complaints that students could not learn after working all day and much perseverance was required to succeed. Nevertheless, technical education was at last recognised by government and, more importantly, it was, by the end of the nineteenth century, state-funded.

Mechanics' Institutes and University Extension Schemes

The idea of university extension courses as a form of adult education evolved during the later 1860s with the idea of establishing local colleges of higher education. The Mechanics' Institute Movement was ideally situated to respond to this and it was no coincidence that most of the support came from the North, where the Lancashire and Cheshire Union and the Yorkshire Union had good reputations for supporting technical education through the examination boards of the Science and Art Departments and Society of Arts. Indeed, the London press referred to those who attended these classes as 'the sturdy artisans of the North' (Marriott 1992: 197).

The idea of offering university courses is attributed to Arthur Henfrey, who wrote an article entitled 'Society of Arts on Industrial Instruction' which was published in the annual report for 1855 (*Seventeenth Annual Report of the Yorkshire Union of Mechanics' Institutes* 1855: 34). Henfrey was born in Aberdeen, and his father was an English engineer. He studied medicine and surgery at St Bartholomew's hospital, London. However, he was unable to continue with a medical career as he suffered from asthma. Henfrey devoted the rest of his life to botany, during which he wrote 39 scientific papers and 11 books, including one entitled *An Elementary Course of Botany*, which was the leading text of the day. His interest in education is apparent through his contributions as a translator and writer of textbooks in relation to science and particularly botany, subjects offered by mechanics' institutes. Henfrey highlighted the importance of having university-trained lecturers for the teaching of higher-level courses in the mechanics' institutes. He believed it was important for mechanics' institutes to be 'converted into colleges' where there was the opportunity for working-class men to receive university-level industrial training. Henfrey believed that if 'university lecturers were examined thoroughly and graduate in each branch [subject], with the view to lecturing on principles in these municipal colleges' then higher education in technical subjects would be available to a large percentage of the working class (Oliver 1913). Henfrey made specific reference to the training of chemists in industrial towns and geologists in mining districts, both of which were supported well within Yorkshire Union institutes (ibid.).

Henfrey not only planned the design of these higher education colleges containing 'a lecture theatre, library and reading room as well as laboratories and studios', but also the sort of curriculum that would be required. For example, he suggested that advanced scientific instruction should be available for those who had already completed previous courses and would last for four years 'in a system of progressive though gradually increasing complexity' (ibid.). In each year the student should attend twenty-four lectures on science, delivered at intervals of a fortnight'. Henfrey even suggested the following scheme of work (see Table 4.2).

Some 30 years later, university extension courses would be offered initially by the established Universities of Oxford and Cambridge, from 1873, and, although lectures were predominately delivered in northern towns, they were also introduced in the East Midlands (Marriott 1992: 197).

Both Universities 'were proud of their northern dependencies; through them they secured a connection with centres of trade and manufacturing', with Oxford predominantly responsible for supporting centres in the textile and engineering districts of Lancashire and the West Riding, and Cambridge concentrated on the North-Eastern circuit, taking in the mining communities, Newcastle and as far north as the Borders (Marriott 1992: 197).

By the 1870s, several institutes of the Yorkshire Union were promoting higher-level education in relation to university extension classes, which were supported by itinerant lecturers (Marriott 1992: 197). Hartlepool Mechanics' Institute was

Table 4.2 Scheme of work

	Subject content
Year 1	Physics (Astronomy, Properties of Matter and Form, Heat, Light, Electricity)
Year 2	Chemistry (Crystallography, Chemical Affinities, Atomic Theory, Mineral Chemistry, Organic Chemistry, Chemistry of Life)
Year 3	Morphology (Vegetable, Animal) and Physiology (Vegetable, Animal)
Year 4	Physical Geography and Geology

Source: Oliver 1913.

anxious that its teachers who had been involved with elementary classes were given the opportunity to gain university extension teaching certificates through Oxford or Cambridge Universities so that they could gain formal recognition by the Education Department to teach subjects at advanced level on the scheme (Jepson 1973: 114).

In 1874, university extension courses were being promoted and delivered at the Keighley Institute on political economy and were 'welcomed as an important educational movement' (*Thirty-Sixth Annual Report of the Yorkshire Union of Mechanics' Institutes* 1874: 134). Darlington, Hartlepool West, Middlesbrough and Stockton Institutes in the North East were associated with a scheme, run through Durham and Cambridge Universities. Courses were offered in political economy, history, mining and geology (Jepson 1973).

Several mechanics' institutes across the Yorkshire Union also offered these examinations, including Barnsley Mechanics' Institute, which introduced them in 1884 through Cambridge University. Twelve lectures were given during the year by W. W. Watts of the Geological Society, and the first lecture, given free, was attended by the mayor of Barnsley. The remaining 11 had a weekly nominal fee. Watts remarked that the university extension movement had been established to serve students of towns that had no university but desired higher education. The work was assessed through questions asked during the lectures and through examinations for which students would gain certificates, if they completed successfully (*Barnsley Chronicle* 4 October 1884).

A series of lectures was presented by academics from Balliol College, Oxford, at the Huddersfield Mechanics' Institute in 1886 and a synopsis of the courses was printed in the *Daily Chronicle* (6 December 1886). The Oxford University extension lectures were successful at Huddersfield in 1888, both in numbers attending, with '400 seated in the large hall', and in the income it received (*Oxford University Extension Lectures Examination Report* November 1886).

Table 4.3 lists the institutes that were part of the University Extension Scheme between 1885 and 1902. The centres associated with the working classes

Table 4.3 Mechanics' Institutes in the Yorkshire Union that were part of the University Extension Scheme

Barnsley	Hartlepool and West Hartlepool	Pontefract	Sunderland
Bradford	Hebden Bridge	Ripon	Thirsk
Cleckheaton	Heckmondwike	Rotherham	Todmorden
Doncaster	Huddersfield	Scarborough	Whitby
Dewsbury	Hull	Sheffield	York and Railway Institute
Doncaster	Ilkley	Shipley	
Filey	Keighley	Skipton	
Halifax	Leeds	Sowerby Bridge	
Harrogate	Middlesbrough	Stockton	

Source: Jepson 1973: 138.

connected with Oxford seem to have been in Lancashire and Yorkshire. It was often the Co-operative Movement that was behind their establishment, and they were in towns that had established institutes.

The institutes at Barnsley, Doncaster, Hebden Bridge, Rotherham, Shipley and Sowerby Bridge laid claim, rightly, to being strictly working class. At Ilkley, Halifax, Cleckheaton, Keighley and Bradford the majority were not working men, but it would be unfair to indicate exclusively that they were not so. Huddersfield Technical School provided university extension courses. By 1892, there were over 50 Oxford centres, and Hebden Bridge, Huddersfield and Ilkley were the most established, but not the only ones in Yorkshire (Jepson 1973: 138). There were of course other centres outside the Yorkshire Union (see Table 4.4).

In 1887, in response to the Samuelson Commission, the Technical Schools (Scotland) Act was passed to support government funding for technical education. Similarly, Scotland introduced university extension courses through Cambridge University (Cooke 2006: 115–9). The Cambridge University Extension Scheme was an attempt to bring university-standard teaching to the public through the mechanics' institute movement, but was privately financed by 'leading citizens' (Roderick and Stephens 1970: 108).

There were also the university extension courses provided by Oxford, London and Durham. It was Hervey, writing in 1855, who suggested the potential of offering education at university standards in the mechanics' institutes when he said: 'On one hand we have a vast number of voluntary literary associations … composed of the middle and working classes, but deficient in the means of obtaining that popular but sound instruction [university education] which they are thirsting for' (Hervey 1855: 10). Hervey went on to say, 'why then might it not be practicable and advisable for universities to supply lecturers to the mechanics' institutes'. He believed that with the developments of a railway network now underway, towns within travelling distance of universities could be provided with professors and good teachers to develop and extend higher-level knowledge to the masses. This was very forward-looking and in keeping with the

Table 4.4 Other university extension centres outside the Yorkshire Union

Accrington	Bury	Manchester	Runcorn
Alderley Edge	Camborne	Matlock	Southampton
Altrincham	Cheltenham	Moston	South Shields
Ancoats	Chester	Nantwich	Stafford
Ashton	Colchester	Oldham	Tunbridge Wells
Bath	Darlaston	Pontypool	Warrington
Birmingham	Gloucester	Pucklechurch	Whitehaven
Bolton	Guildford	Reading	Winton
Bournemouth	Hove	Redruth	Workington
Brighton	Hyde	Rhyl	Co-operative Society
Bristol	Lewes	Rochdale	

Source: Jepson 1973: 164–165.

growing relationship between higher and further education in the twenty-first century.

The University Extension Scheme lectures were offered in many of the larger towns that had accommodation for the events to take place, which mechanics' institutes had. Nottingham was the first town to participate in the scheme, followed by others quite soon afterwards. The University Extension Scheme 'increased the usefulness of universities by bringing them into touch with a wider circle of students'. For example, 'in Northumberland, between 1879–1887 University Extension Lectures appealed to the rank and file of a typical industrial district' (Wiltshire, Taylor *et al.* 1980: 28).

In the North-East cluster several institutes were promoting higher-level education by the 1870s. Hartlepool elementary teachers were one group who were particularly keen on gaining university extension teaching certificates, through Oxford and Cambridge, so that they could gain recognition through the Education Department (Jepson 1973: 114). At Darlington in 1880, the Durham University extension lectures were held in the Institute's 'excellent Lecture Room' (*Darlington Mechanics' Institute Annual Report* 1880: 7).

Although it is difficult to measure to what extent the University Extension Scheme had an effect academically, it does strongly indicate that there was a demand for adult scientific and technical education. Nicholas (1985: 85) summarises thus: 'It was unlikely that the extensive system of evening and part-time science and technical instruction in Britain could have been developed without building on the foundations of the mechanics' institutes.' Durham, Cambridge, Oxford and London Universities provided lectures in provincial mechanics' institutes that were the forerunners of degree-level courses for adults established in the twentieth century. Thus, mechanics' institutes were also responding to higher levels of technical education, some at degree level, through the University Extension Scheme. The fact that the traditional Universities of Oxford, Cambridge and later Durham and London, were involved with the mechanics' institute movement in this way suggests they were confident that the teaching was being delivered at a high level in the provinces.

Thus, the Mechanics' Institute Movement had adapted in various ways to support scientific and technical education during the second half on the nineteenth century and, while there were economic challenges to Britain's economic supremacy, mechanics' institutes had provided a foundation on which twentieth-century further and higher education could be developed. Curtis (1967: 473) argues that 'mechanics' institutes were … an important step in the development of scientific and technical instruction'. He identified that some institutes in Birmingham, Leeds, London and Manchester developed into technical institutes and colleges. While the London Mechanics' Institute evolved into Birkbeck College and later part of the University of London, others, such as those at Bradford, Huddersfield and Leeds and at other important towns, became technical colleges. Some, notably Huddersfield, used the former mechanics' institute as the first building, while others were established in new turn-of-the-century purpose-built technical schools and colleges. As Fieldhouse (1998: 46) succinctly puts it:

The [twentieth] century began with a flurry of educational activities, the first of which was the passing of the 1902 Education Act. This attempted to bring order to the imbroglio of public education that had emerged during the nineteenth century.

The 1902 Balfour Education Act required Local Education Authorities, through local rates, to take responsibility and manage elementary, secondary and adult education. Local technical committees were to be set up to manage what would become technical and further education colleges, many previously having been mechanics' institutes. Examples, with their known name in 1956, include Heriot-Watt College (1821), Lancaster and Morecambe College of Further Education (1824), Leeds College of Technology (1824), and Manchester Municipal College (1824), Huddersfield Technical College (1825, reopened 1841), Dudley and Staffordshire (1862) and Cardiff College of Technology and Commerce (1865). Later nineteenth-century institutions, prior to the passing of the 1902 Act and having connections in their local mechanics' institutes, include the City of Gloucester Technical College (1873), Birmingham College of Technology (1895), Leicester College of Art (1896), Brighton Technical College (1897) and Stretford Technical College (1899) (Venables 1956: 49–62).

Conclusion

As Harrison (1961: 213) notes, 'from educational bankruptcy the mechanics' institutes were rescued by two timely developments – the examination system of the Society of Arts, and the demand for technical and scientific education'. Even as late as the 1860s and 1870s the phrase 'technical education' seems to have been rarely used and instead the common term was 'scientific instruction'. With the Samuelson Commission, technical instruction became the more common term for both specific and general adult education, the former supporting practical instruction for industry and the latter for those gaining some elementary education (Musgrave 1964). By the late 1880s, many of the larger mechanics' institutes offering examinations were in a strong position to become technical schools, such as those at Bradford, Bristol, Edinburgh, Glasgow, Huddersfield, Keighley, Leicester, Manchester, Oldham and Preston (Harrison 1961: 151). Such institutions, many now with purpose-built accommodation and some still having mechanics' institutes in their title (see Chapter 7, this volume), were developing reputations that were nationally recognised for providing adult technical education. Internationally, there was some way to go. Cruickshank (1974: 152) puts it succinctly when he says: 'In Germany, for example, the city of Berlin had recently [1880s] spent £300,000 on a school for mechanical engineering, and even the Saxon town of Chemnitz, no larger than Oldham, had eight technical schools.'

Crucially, with the formulation of reputable examinations in technical subjects offered by the Science and Art Department, Society of Arts, City and Guilds London Institute and several others, and the passing of the Technical Instruction

Act in 1889 and the Local Taxation (Custom and Excise) Act of 1890, an injection of public money through new local authorities established under the 1888 County Councils Act finally put adult education of a sound financial footing. Together, these Acts contributed to what Green (1995: 138) has referred to as 'the golden age of the English technical education movement'.

Note

1 Huddersfield Institute became the northern centre for conducting Society of Arts Examinations in 1857. It was selected for its reputation in administrating other external examinations and good teaching. The following year, other centres did the same.

References

Argles, M. (1964) *South Kensington to Robbins, An Account of English Technical and Scientific Education since 1851* (London: Longmans).

Auerbach, J. A. (1999) *The Great Exhibition of 1851, A Nation on Display* (Yale, Connecticut: Yale University Press).

Baker, G. (2001) 'The romantic and radical nature of the 1870 Education Act', *History of Education* Vol. 30, No. 3, 224–226.

Barnsley Chronicle (4 October 1884) (unknown author).

Barnsley Chronicle (6 December 1886) (unknown author).

Barton, S. (2000) 'Why should working men visit the Exhibition?: Workers and the Great Exhibition and the ethos of industrialism', Inkster, I., Griffin, C., et al. (eds) *The Golden Age, Essays in British Social and Economic History, 1850–1870* (Aldershot, Hampshire: Ashgate): 144–163.

Betts, R. (2006) 'Persistent but misguided? The technical educationalists 1867–89', *History of Education: Journal of the History of Education Society* Vol. 27, No. 3, 267–277.

City and Guilds of London Institute (1993) *City and Guilds of London Institute: A Short History, 1878–1992* (London: City and Guilds of London Institute).

Cooke, A. (2006) *From Popular Enlightenment to Lifelong Learning* (Leicester: National Institute of Adult Continuing Education).

Cruickshank, M. J. (1974) 'From Mechanics' Institution to Technical School', Cardwell, D. S. L. (ed.) *Artisan to Graduate, Essays to Commemorate the Foundation in 1824 of the Manchester Mechanics' Institution, Now in 1974 the University of Manchester Institute of Science and Technology* (Manchester: Manchester University Press): 135–156.

Curtis, S. J. (1967) *History of Education in Great Britain* (London: University Tutorial Press).

Darlington Mechanics' Institute (1880) *Annual Report*.

Davis, J. R. (1999) *The Great Exhibition* (Stroud: Sutton Publishing).

Dawes, R. (1856) *Mechanics' Institutes and Popular Education, An Address Delivered at the Annual Soirée of the Huddersfield Institute, December 13 1855* (London: Groombridge and Sons).

Dunstan, D. (1996) *Victorian Icon: The Royal Exhibition Building Melbourne* (Melbourne: The Exhibition Trustees in Association with Australian Scholarly Publishing).

Elliott, P. (2004) 'Improvement, always and everywhere: William George Spencer (1790–1866) and the mathematical, geographical and scientific education in nineteenth-century England', *History of Education: Journal of the History of Education Society* Vol. 33, No. 4, 391–417.

Fieldhouse, R. (1998) *A History of Modern British Adult Education* (Leicester: National Institute of Adult Continuing Education).

Garner, A. D. (1985) 'The Society of Arts and the mechanics' institutes: The co-ordination of endeavour towards scientific and technical education, 1851–54', *History of Education: Journal of the History of Education Society* Vol. 14, No. 4, 255–262.

Green, A. (1995) 'Technical education and the state formation in nineteenth-century England and France', *History of Education: Journal of the History of Education Society* Vol. 24, No. 2, 123–139.

Gurney, P. (2001) 'An appropriated space: The Great Exhibition, the Crystal Palace and the working class', Purbrick, L. (ed.) *The Great Exhibition of 1851, New Interdisciplinary Essays* (Manchester: Manchester University Press): 114–45.

Harrison, J. F. C. (1961) *Learning and Living, 1790–1960: A Study in the History of the English Adult Education Movement* (London: Routledge and Kegan Paul).

Hemming, J. P. (1978) 'Some attempts at commercial education in the Mechanics' Institutes', *The Vocational Aspect of Education* Vol. XXX, No. 75, 41–44.

Hervey, A. C. (1855) *A Suggestion for Supplying the Literary, Scientific and Mechanics' Institutes of Great Britain and Ireland* (Cambridge: Macmillan and Co.).

Hobhouse, H. (2002) *The Crystal Palace and the Great Exhibition, Art, Science and Productive Industry* (London: Athlone).

Hole, J. (1853) *An Essay on the History and Management of Literary, Scientific, and Mechanics' Institutions* (Reprint. London: Cass 1970).

Huddersfield Mechanics' Institute (25 November 1857) *Minutes of the General Committee* (Huddersfield: University of Huddersfield archives).

Jepson, N. A. (1973) *The Beginnings of English University Adult Education* (London: Michael Joseph).

Lancashire and Cheshire Union of Mechanics' Institutes (1891) *Annual Report.*

Maclure, J. S. (1969) *Educational Documents, England and Wales 1816–1968* (Trowbridge, Wiltshire: Redwood Press).

Manchester Guardian (8 February 1862) Annual Meeting of the Lancashire and Cheshire Union (unknown author).

Marriott, S. (1992) 'University extension in the North of England and the Leeds historians', *Northern History* Vol. XXVIII, 197.

Marsden, W. E. (Summer 1969) 'The growth of technical education in Southport, 1874–1944', *The Vocational Aspect of Education* Vol. XII, No. 24, 44–62.

Musgrave, P. W. (1964) 'The definition of technical education: 1860–1910', *The Vocational Aspect of Education* Vol. 16, No. 34, 105–111.

Nicholas, S. J. (1985) 'Technical education and the decline of Britain 1870–1914', Inkster, I. (ed.) *Steam Intellect Societies: Essays on Culture, Education and Industry, circa 1820–1914* (Nottingham: University of Nottingham): 45–85.

Northern Counties Technical Examinations Council (1945) *Northern Counties Technical Examinations Council 1920–1945: A Record of Twenty-Five Years* (Newcastle-upon-Tyne: Mawson, Swan and Morgan Ltd).

Northern Counties Technical Examinations Council (1970) *Northern Counties Technical Examinations Council 1920–1970: A Record of Fifty Years* (Newcastle-upon-Tyne: Mawson, Swan and Morgan Ltd).

Oliver, F. W. (ed.) (1913) 'Arthur Henfrey, maker of British botany', *The Oxford Dictionary of National Biography*. Available online at http://www.oxforddnb.com/index/101012922/Arthur-Henfrey.

Oxford University Extension Lectures Examination Report (November 1886) (Huddersfield: University of Huddersfield archives).

Parry, J. (2015) 'Spencer Compton Cavendish', *The Oxford Dictionary of National Biographies*. Available online at www.oxforddnb.com/index/101032331/Spencer-Cavendish.

Roderick, G. W., and Stephens, M. D. (1970) 'Approaches to technical education in nineteenth-century England', *The Vocational Aspect of Education* Vol. XXII, No. 52, 103–111.

Royle, E. (1987) *Modern Britain, A Social History, 1750–1985* (London: Hodder Arnold).

Smith, S. (1886) *Night Schools and Technical Education* (Leeds).

Royal Society of Arts (1852–1856) *Royal Society of Arts Manuscript Subscription Books* (London: Royal Society of Arts archives).

Stephens, M. D., and Roderick, G. W. (1972) 'Science, the working class and Mechanics' Institutes', *Annals of Science* Vol. 29, No. 4, 349–360.

Tylecote, M. (1957) *The Mechanics' Institutes of Lancashire and Yorkshire Before 1851* (Manchester: Manchester University Press).

Venables, P. F. R. (1956) *Technical Education: Its Aims Organisation and Future Development* (London: Bell and Son).

Walker, M. (2008) *Examinations for the 'underprivileged' in Victorian times: The Huddersfield Mechanics' Institution and the Society for the encouragement of Arts, Manufactures and Commerce* (London: William Shipley Group).

Walker, M. (2013) 'Encouragement of sound education amongst the industrial classes: Mechanics' Institutes and working-class membership 1838–1881', *Educational Studies* Vol. 39, No. 2, 142–155.

Walker, M. (2015) 'The impact of the Great Exhibition of 1851 on the development of technical education during the second half of the nineteenth century', *Research in Post Compulsory Education and Training* Vol. 20, No. 2, 193–207.

Ward, L. O. (1973) 'Technical education and the politicians 1870–1918', *British Journal of Educational Studies* Vol. 21, No. 1, 34–39.

West Riding of Yorkshire Union of Mechanics' Institutes (1838) *First Annual Report.*

Wiltshire, H., Taylor, J., and Jennings, B. (eds) (1980) *The 1919 Report. The Final and Interim Reports of the Adult Education Committee of the Ministry of Reconstruction 1918–1919* (Nottingham: University of Nottingham).

Wright, G. (2001) 'Discussions of the characteristics of mechanics' institutes in the second half of the nineteenth century: The Bradford example', *Journal of Educational Administration and History* Vol. 33, No. 1, 1–16.

Yorkshire Union of Mechanics' Institutes (1844) *Sixth Annual Report.*

Yorkshire Union of Mechanics' Institutes (1849) *Eleventh Annual Report.*

Yorkshire Union of Mechanics' Institutes (1851) *Thirteenth Annual Report.*

Yorkshire Union of Mechanics' Institutes (1855) *Seventeenth Annual Report.*

Yorkshire Union of Mechanics' Institutes (1857) *Nineteenth Annual Report.*

Yorkshire Union of Mechanics' Institutes (1874) *Thirty-Sixth Annual Report.*

Yorkshire Union of Mechanics' Institutes (1880) *Forty-Second Annual Report.*

Yorkshire Union of Mechanics' Institutes (1882) *Forty-Fourth Annual Report.*

Yorkshire Union of Mechanics' Institutes (1891) *Fifty-Third Annual Report.*

5 Social class and membership

Introduction

Elizabeth Gaskell in her novel, *Mary Barton, A Tale of Manchester Life*, published in 1848, highlighted the enthusiasm amongst weavers for learning.

> There is a class of men in Manchester, but they are scattered all over the manufacturing districts of Lancashire ... in the neighbourhood of Oldham there are weavers, common hand-loom weavers, who throw the shuttle with unceasing sound, though Newton's *Principia* lies open on the loom, to be snatched at in work hours, but revelled over in meal times, or at night. Mathematical problems are received with interest, and studied with absorbing attention by many a broad-spoken common-looking factory-hand (Gaskell 1967: 34).

Some 47 years later, in his play, *The Importance of Being Earnest*, Oscar Wilde's character Lady Bracknell reminds the reader in Act 1 that, 'fortunately in England at any rate, education produces no effect whatsoever; if it did, it would prove a serious danger to the upper classes and probably lead to acts of violence in Grosvenor Square' (Wilde 1998: 265).

Historians have argued that the membership of mechanics' institutes was made up of members of the middle class and professional classes, among them Royle (1971) and Luckhurst (1959), which is a valid general observation. However, for institutes in the northern counties at least, this was not always the case from about the mid-nineteenth century onwards. Historians' assumptions were based on research carried out up to the 1850s, when it was believed the movement had reached its zenith and was seen as declining in its importance, having not achieved its initial aim of offering relevant education to the working class (Hudson 1851: 41, Inkster 1985: 123).

The term 'class' emerged only in the 1820s after large-scale economic and social changes had taken place in the late eighteenth and early nineteenth centuries. Before industrialisation, written accounts made reference to social hierarchy as 'ranks', 'orders' and 'common people'. Briggs (1960) states that by 1824 the word 'class' had been added to the vocabulary, replacing these terms. Thus, at the time that mechanics' institutes were being established throughout the country, the term 'class' was already being used.

Early developments

Many mechanics' institutes committees did not go out of their way to exclude anyone, male or female, and they were keen to accept members from a cross-section of society. For example, in their first address in 1825, the directors of the then recently opened Huddersfield Mechanics' Institute stated that 'the great object of this Institution is to bring within the research of all, but more particularly the trading and working classes, the acquisition of useful knowledge', a common objective of many such institutions (*Rules of the Huddersfield Scientific and Mechanic Institute, for the Promotion of Useful Knowledge* 1825: 1). The address is evidence that from their early origins, mechanics' institute committees believed that education should be within the reach of everyone and, in particular, the newly evolving labouring population, at a time when there were rapid and far-reaching social and technological changes taking place.

As previously mentioned, the choice of name, 'mechanic', was no coincidence. Other educational establishments including philosophical and scientific and literacy societies had been established before mechanics' institutes, and their society names were in no way connected with the working classes. Mechanics' institutes, however, at their formation in the 1820s made every effort to appeal to the wider population through naming them as such, without realising that the name discouraged working-class men from attending with mechanics, who might be skilled workers reluctant to attend in case their expertise and knowledge were 'stolen' by others, or non-skilled labourers on the grounds that mechanics' institutes were not for them but for skilled mechanics. This reinforces the belief that the nineteenth-century skilled workers often differentiated themselves from the rest of the working class, insisting on having honour and respect for their abilities. As McWilliam (1998: 53) states, 'the working classes did not always think in terms of natural class alliances'. Therefore, it is not surprising that both contemporaries, such as Hudson (1851), and historians have confidently claimed that as neither the 'mechanic' nor the working class generally attended these institutes, which were set up especially for them, the movement 'failed'.

Not surprisingly, the institutes were financially supported by the middle and upper classes and, therefore, offered education which benefactors thought relevant to the whole population (Langley 1849: 321). As a result, of 204 mechanics' institutes in England and Wales in 1849 only 43 were largely supported by operatives and mechanics. The Manchester Mechanics' Institute, for example, was 'beyond the reach of the great manufacturing population' and this was very much the case in other towns (Hudson 1992: 144). The same was also true of Liverpool and London. Of 32 institutes in Lancashire and Cheshire, only four were attended by 'considerable numbers' of the working classes, and of the 21 institutes in the Midlands, only three had such members (Royle 1971: 305). Robert Elliot summed up the situation in 1861, as he perceived it, when he wrote, 'the banquet was prepared for guests who did not come' (Elliot 1861: 676).

Thus, the findings of contemporaries with regard to mechanics' institutes between 1824 and the 1850s support the view held by many historians that the majority of members attending them were from the middling and professional classes. Luckhurst states categorically that:

> Mechanics' institutes ceased to deserve their distinctive name as so few artisans were sufficiently well educated to profit from the classes, lectures, libraries and other educational facilities which they provided, and their places were soon filled by clerks and office workers, whose numbers were greatly increasing through the rapid expansion of commerce (Luckhurst 1957 Chapter X: 4).

As previously identified, many institutes adapted their subject classes to be more relevant to industrialisation than scientific experimentation in order to attract the 'gentling masses'. To address this point further, that the failure of the mechanics' institutes was in not attracting the working classes, it is necessary to consider the question of class in some detail.

The term 'class' was being used more commonly from the 1830s, at the time when the Mechanics' Institute Movement was expanding, identifying the relationships between different social groups. Karl Marx argued that when 'millions of families live under economic conditions which separate their way of life, their interests and their education from those of other classes, they constitute a class in itself'. For him, class was about relationship to the means of production – it was about source of income and it determined culture (Hudson 1992: 202).

In 1832, the failure of the Reform Act to enfranchise those who considered themselves working class, the opposition to the Poor Law Amendment Act of 1834, and social problems, such as bad housing and sanitation and overcrowding, had much impact on class formation (McWilliam 1998). It was these kinds of events which, as E. P. Thompson (1968: 8) argued, brought men with similar or common experiences together to articulate their identity of interests between themselves, and which they saw as different from other groups, which 'ties loosely together a bundle of discrete phenomena. There were tailors here and weavers there, and together they made up the working classes'. Thompson argued that class-consciousness was essential, that class was not just economic but cultural too. He believed that 'class happens when some men, as a result of common experiences, inherited or shared, feel and articulate the identity of their interests as between themselves, and against other men whose interests are different from theirs' (ibid.: 13).

Thompson argued that the period from the 1790s to the 1830s saw the making of the English working class. In the context of this work, he believed that there was a growing identity of interests between diverse groups of the working population, which were distinct from other classes, 'strongly based and self-conscious working-class institutions and movements; friendly societies, trade unions, educational and religious movements, co-operatives, working-class periodicals and Chartism' (Hudson 1992: 142). Thompson (1968: 23) identified certain occupations, such as 'stockingers, handloom weavers, cotton spinners, artisans, shoemakers, small masters, tradesmen, publicans, book sellers and professional men' that became the main thrust of this self-conscious working-class institutional movement, the same occupations that often contributed to the membership of mechanics' institutes.

Harris (1993: 148) states that the unifying of class consciousness was not so much through the fact that 80 per cent of the population was made up the working class but did not see themselves as such, but instead 'through ideology, political movements and in leadership of minority organisations'. Such minority organisations included trade unions, cooperatives and mechanics' institutes or similar.

However, the working class was never as united or as self-confident as Thompson suggested. This social group was more fragmented, with marked divisions between urban and rural, skilled and unskilled workers. Perkin (1986: 264) cites an unreferenced contemporary account that supports this.

> The ploughmen hold the mechanics in contempt as an inferior race of beings, although the latter can earn the best wages: The journeymen cabinet makers cannot degrade themselves by associating with the journeymen tailors, the journeymen shoemakers cannot so forget their dignity as to make companions of the labourers.

It is therefore not surprising that during the first 20 years or so that mechanics' institutes were being established, the working class often had no allegiance to them or what they were offering despite their committees believing that lectures and subjects were relevant to industry. Few working-class adults had had an elementary education, as previously identified.

The largest group of the Victorian middle class was the one associated with the professions and it was this group who particularly supported the mechanics' institutes and working-class education. During the first half of the nineteenth century entrepreneurship increased the numbers of employees who became part of this class with a background in trade and business (Tosh 1999) – men such as Isaac Holden, the son of a pit headsman, who became a successful woollen master in Bradford. He had started as a piecer at the age of ten, and by the time he was 40, he was the owner of his own worsted mill and was known as the 'first comber in Europe'. He became immensely wealthy with an aristocratic-style mansion, a seat in the House of Commons and a knighthood. The newly developing industrialists, bankers and merchants became part of the backbone of the 'middling' middle class. They were distinguished from the aristocracy and gentry because they worked for a living, and were originally from the working class (Perkin 1986).

The membership of mechanics' institutes initially reflected this change in class order. Presidents of institutes were often from long-established aristocratic families, such as Sir John Ramsden, Baronet, at Huddersfield, and the Dukes of Devonshire, who supported several institutes, including those at Grassington and

Keighley in the West Riding of Yorkshire. The seventh Duke was the Chair of the *Report of the Royal Commission on Scientific Instruction and the Advancement of Science*, published in 1875. However, as mechanics' institutes began to reinvent themselves from the 1840s, presidents often came from the new middle class, such as Frederic Schwann, who re-established the Huddersfield Institute in 1841.

Crucially, in relation to this work, the language of class highlights that the middle class and working class, or at least those who were seen as 'Radicals', had more in common than was previously thought. While both had their own class identity, they also had much in common. Middle-class radicals supported the establishment and running of voluntary organisations for all, having a deep mistrust of state intervention. Working-class Radicals aligned themselves with middle-class sympathisers in relation to politics and 'self-help' during the nineteenth century. The mechanics' institutes are a good example of this, with middle-class Radicals, such as Schwann, a local export merchant of textiles who opposed the Corn Laws, supporting the working class (his employees) with 'self-improvement' as President of the Huddersfield Mechanics' Institute Committee.

The debate on class

Hobsbawm (1964) conducted substantial and highly acclaimed work on class. He put the working class into three distinct categories over three defined periods; the first period (1780s to 1815) was 'the classical age of the Industrial Revolution [which] saw the birth of the modern working class' and the period prior to the foundation of the mechanics' institute movement. The second period (1840s to 1870s) saw the establishment of capitalism, which 'ruled supreme', and labour aristocracy developed, the same period during which the major expansion of mechanics' institutes also took place. The third period (1890s to 1914) was one of imperialism, 'monopoly capitalism' and mass production and expansion of secondary and tertiary industries, the period during which the state established and funded both school-age and adult education. The latter included technical colleges and schools of art, and many of them had been previously mechanics' institutes following the expansion of technical education (Hobsbawm 1964: 272).

Hobsbawm (1964) stipulates that there were six specific factors that should be considered in relation to labour aristocracy, that is, those adults who were in regular paid work and had generally good living conditions and status between the 1840s and 1870s. These are listed in Table 5.1.

Those at the top of the working-class social strata tended to merge with the 'lower middle class' and, in the first half of the nineteenth century, this group would include small shopkeepers, foremen and managers, the latter being promoted former workers. By the end of the century, clerks, bookkeepers, managers 'and the better sort of working folk' (Hobsbawm 1964: 273) as distinct from employers, solicitors, physicians and tradesmen, were on the line between being lower middle class and upper working class, who Hobsbawm refers to as the 'labour aristocrats' (ibid.).

Table 5.1 Hobsbawm's six factors relating to the labour aristocracy

Factor	Criterion
1	Level and regularity of a worker's earnings.
2	His prospect of social security.
3	His conditions of work, including the way he is treated by foremen and masters.
4	His relations with the social strata above and below.
5	His general condition of living.
6	His prospects of future advancement and those of his children.

Source: Hobsbawm 1964: 273.

It was the Victorians who were the first to refer to manual workers 'as the aristocracy of the labouring classes' or the upper working class (Hobsbawm 1984: 183). Contemporary middle-class observers thought that they made up about 50 per cent of all manual workers but Hobsbawm believes that they were a somewhat smaller group. He identified that this group were superior in terms of their economic status such as having higher, regular wages and who had the opportunity to save some of their income. Their social, political and cultural standing in society was also an indicator. They were seen as respectable and moral. The working-class aristocracy was associated with various organisations such as trade unions, cooperatives and friendly societies; although he makes no reference to mechanics' institutes specifically, the same must have been true (ibid.: 227).

During the period of the Great Depression (1873 to 1896) there was a growth in class-consciousness as a result of the tensions during the period. There was also the rise in tertiary employment and a rise in real wages due to falling living costs during the 1880s. Both the lower middle class and upper working class were attempting to 'better themselves' by personal effort and self-help, which both subscribed to. For 'one group [the working class] self-help could not become real without collective institutions such as friendly societies and cooperatives' (ibid.: 243). Mechanics' institutes should be included in the list of institutions that attracted the upper working class and lower working class, as evidenced by several institutes' lists of occupations.

Hobsbawm's third period covers 30 years during which class identity changed. The skills of the workforce continued to develop and resulted in exploitation between the skilled or 'privileged workers [who] systematically protected their position by exploiting weaker [non-skilled] workers' (Harris 1993: 10). In spinning mills, for example, 'minders' deliberately excluded the 'piecers', who they saw as inferior, from the acquisition of new skills, which resulted in preventing them from progressing in their jobs as well as controlling their wages and even compelling them to join a separate union, subscriptions to which were deducted by the minders from the piecer's weekly wage. This may well account for the fact that few piecers were mentioned in lists of occupations that attended mechanics' institutes during the first half of the nineteenth century (ibid.).

The example of piecers indicates, however, that there was a changing nature with regards to occupational membership of mechanics' institutes from the mid-nineteenth century onwards. In the case of mechanics' institutes of the Yorkshire Union up to 1850, only Todmorden listed 'piecers' as one of several occupations, of whom there was only one attending in 1841 and two in 1843. Neither Keighley nor Huddersfield had listed them. However, 'piecers' were well represented amongst the membership during the second half of the nineteenth century and this is likely to be due to the institutes being more welcoming to the working class generally and the less skilled, or lower working class, in particular (ibid.). Many institutes also reduced membership fees and offered fortnightly, rather than quarterly, payments for flexibility during periods of depression, when members were unable to pay, or during periods of prosperity when employers expected them to work longer hours. In the case at Huddersfield in 1857, membership included 47 piecers out of a total membership of 662, or just over 7 per cent (see Table 5.10).

Neale (1981) has sought to categorise nineteenth-century class with a five-class model of social structure. Like Hobsbawm's work, this is very useful, for it allows the use of occupation lists of members of mechanics' institutes to be further considered in relation to the precise social status of their membership (see Table 5.2).

Neale's five-class model provides useful categories that help identify class and membership occupations of those who attended mechanics' institutes and, crucially, provides further evidence that the working class began to be attracted to them, often supported by the middle class, particularly but not exclusively the Radicals. Neale's model summarises occupations in relation to class that can be related to occupation of members as listed by institutes in their Annual Reports. The Manchester Mechanics' Institute, for example, was associated during the

Table 5.2 Neale's five-class model

Class	Component
Upper class	Aristocratic, landowning, authoritarian, exclusive.
Middle class	Large industrial and commercial property-owners, senior military and professional men aspiring to acceptance by the upper class.
Middling class	Petit bourgeois. Aspiring professional men, other literates and artisans, concerned about removing the privileges and authority of the upper class in which, without radical changes, they cannot realistically hope to share.
Working class 'A'	The industrial proletariat in factory areas, workers in domestic industries, collectivist and non-deferential and wanting government intervention to protect rather than liberate them.
Working class 'B'	Agricultural labourers, other low-paid non-factory urban labourers, domestic servants, urban poor, most working-class women whether working class 'A' or 'B' households.

Source: Neale 1981: 133.

1820s with delivering scientific lectures to the professional classes, such as the one John Davis addressed to over 1000 people in his first lecture on natural philosophy in 1829 (Inkster 1975: 456). However, within two years there had been change in membership occupations, as early as 1831, when out of 330 members the vast majority were working class. Table 5.3 shows the breakdown by occupation.

The Ancoats Mechanics' Institute in Manchester, on the other hand, from its foundation in 1838, concentrated on offering elementary education. In 1842 there were 310 members of whom 210 were in occupations known to be working class, among them carders, weavers and spinners. Of the remaining 100, the occupations of 80 were not working class and 20 were not recorded (*Report from the Select Committee on Education, Manchester and Salford* 1853: 74–8).

The Select Committee on Public Libraries of 1849 collected various data including that on the 'class of persons' using institutions, and from this Stockdale (1994: 255) noted that, at both Darlington and Morpeth Institutes, membership was made up of tradesmen and mechanics, while at Middlesbrough, Newcastle and Sunderland, the majority were tradesmen who attended. At Hexham they were from all classes and at Shildon both pitmen and mechanics attended the Institute. The occupations therefore seem to reflect the local catchment area and relevance to industry.

The same was also true for Bradford in 1838, when the Committee reported to the newly formed Yorkshire Union that out of 540 members, 300 were from 'the class for whom especial benefit it is intended to support', namely, the working class (*First Annual Report of the Yorkshire Union of Mechanics' Institutes* 1838: 19). In the 1860s, John Godwin studied the occupations of fathers and sons at Bradford Mechanics' Institute. The majority, he found, were from the working class (see Table 5.4). Godwin's work strongly supports the view that the

Table 5.3 Known occupations of the membership at Manchester Mechanics' Institute 1831

Occupation	Number	Occupation	Number	Occupation	Number
Accountants	3	Dyers, drysalters	2	Plumbers, glaziers	8
Architects, surveyors	5	Gas-lighter	1	Solicitor	1
Bookmaker	1	Gentlemen	5	Surgeon	1
Bricklayer	1	Joiners, builders	22	Tailors	5
Cabinetmakers	6	Hairdressers	3	Tin plate worker	1
Carders	5	Mechanics, machinists	43	Tool-makers	2
Carrier	1	Manufacturers, merchants	7	Warehousemen	55
Clerks	83	Packers	2	Warpers, weavers	9
Cotton-spinners	12	Painters	3	Youths	43
Total					330

Source: Tylecote 1957: 237.

Table 5.4 Known occupations of fathers and sons at Bradford Mechanics' Institute 1842

Fathers' occupation	Number	Sons' occupation	Number
Unknown	6	Wool-comber	1
Wool-sorters	6	Factory boys	4
Wool-combers	5	Shop boy	1
Plasterer	1	In warehouses	10
Sawyer	1	Apprenticed to mechanics	3
Overlookers	3	Shoemaker	1
Weaver	1	Printer	1
Middle class	3	Joiner	1
Working in offices	3	Bookbinder	1
Total	29	Total	23

Source: Godwin 1859: 342.

Table 5.5 Percentage of members per occupation at Bradford Mechanics' Institute 1859

Occupations	Percentage (%)
Life members: merchants, manufacturers, etc.	12
Ministers, professional men, schoolmasters, etc.	5
Shopkeepers, bookkeepers, agents, cashiers, etc.	18
Mechanics, labourers, warehousemen, etc.	65
Total	100

Source: Godwin 1860: 343.

working-class aspirations of fathers were being passed on to their sons in order that they too could 'better themselves'.

Godwin (1860: 344) noted in 1859 that

> on the broad platform of Mechanics' Institutes men of all ranks and opinions have united to assert equality of intellectual rights. It may be said with equal truth, that, next to Sunday schools, they have been one of the strongest moving powers in the work of popular education, hitherto the greatest work of the nineteenth century'.

He identified that membership patterns identified that mechanics, labourers, warehousemen and others made up the highest percentage of members (see Table 5.5).

Godwin (ibid.: 343) reported that the Bradford Mechanics' Institute had seen

> an unbroken stream of youths, sons of working men, rising to the positions of responsibility, which in all probability they never would have filled without its [Bradford Mechanics' Institute's] aid, and in many cases entering upon and pursuing a successful middle-class career by the habits, the knowledge, and connexions acquired in this Institute.

Godwin's work indicates that working-class members were attending the Institute and shows social and economic mobility through education.

At the Keighley Mechanics' Institute in 1840, for example, membership included three surgeons and two solicitors, who are classified by Neale as being middle class. However, there were also 14 spinners, six wool-sorters, three wool-combers and three overlookers, who were from his category of working class 'A' (see Table 5.6).

At the Todmorden Mechanics' Institute in 1841, there were few, if any, occupations that would be defined by Neale as from the middling class. This is not surprising, when the population of the town was smaller than that of Keighley and, with the exception of one or two factory-owning families, it was a typical Pennine working-class textile community. Working class 'A' members included 13 weavers, six overlookers, three mule spinners and three warehousemen (see Table 5.7).

In 1843, there were two middle-class solicitors and a surgeon but the majority continued to be working-class occupations, including four warehousemen, three overlookers and three weavers from working class 'A' (see Table 5.8).

Huddersfield Mechanics' Institute was opened in 1825 but closed due to a small membership. It reopened again in 1841 and in 1844 'nearly the whole of the members are operatives in the receipt of weekly wages. The ages of members vary from 12 to 50; the greater part from 12 to 24' (*Sixth Annual Report of the Yorkshire Union* 1844: 40). In 1847 the occupations of members at the Huddersfield Mechanics' Institute were substantially working class (see Table 5.9).

The Huddersfield Committee reported in 1848 that 'the members of the Institution belong almost exclusively to the working classes' and it particularly, and rightly, admired the fact that 'after their day's labour is over in the workshops and factories, they come hither to seek instruction, instead of spending their time

Table 5.6 Known occupations of the membership at Keighley Mechanics' Institute 1840

Occupation	Number	Occupation	Number	Occupation	Number
Butcher	1	Joiners, cabinetmakers	3	Surgeons	3
Bookkeepers, bankers	5	Linen drapers	3	Schoolmasters	4
Corn dealers	2	Licensed victualler	1	Spinners, manufacturers	14
Clergy, gentry	8	Mechanics, moulders	5	Solicitors	2
Clogger	1	Wool-sorters	6	Cordwainer	1
Masons	3	Watchmaker	1	Druggists	4
Overlookers	3	Wool-combers	3	Farmers	4
Painters, gilders	2	Grocers	2	Plasterer	1
Hatter	1	Printers, stationers	2	Hairdresser	1
Reed makers	2	Ironmonger	1	No trade	5
Total					94

Source: *Second Annual Report of the Yorkshire Union of Mechanics' Institutes* 1840: 87.

Table 5.7 Known occupations of the membership at Todmorden Mechanics' Institute 1841

Occupation	Number	Occupation	Number	Occupation	Number
Builder	1	Moulders	3	Shoemaker	1
Corn dealers	2	Mule spinners	3	Stripper	1
Druggist	1	Mechanic	1	Sizer	1
Farmer	1	Overlookers	6	Solicitor	1
Grocers	4	Printers	3	Weavers	13
Ironmonger	1	Picker-maker	1	Warehousemen	3
Ironfounder	1	Piecers	1	Whitesmiths	2
Junior	1	Rope-maker	1	Other	14
Merchants	4	Schoolmasters	3		
Manufacturers	4	Schoolboy	1		
Total					79

Source: *Third Annual Report of the Yorkshire Union of Mechanics' Institutes* 1841: 28.

Table 5.8 Known occupations of the membership at Todmorden Mechanics' Institute 1843

Occupation	Number	Occupation	Number	Occupation	Number
Bookseller	1	Manufacturers	2	Solicitors	2
Builder	1	Mechanic	1	Schoolmasters	3
Clog- and pattern-makers	2	Moulders	2	Surgeon	1
Druggist	1	Mule spinners	2	Sizer	1
Dissenting Minister	1	Overlookers	3	Warehousemen	4
Hardwareman	1	Printer	1	Weavers	3
Ironfounder	1	Painter	1	No trade or profession	9
Land surveyor	1	Piecers	2		
Merchants	4	Rope-maker	1		
Total					51

Source: *Fifth Annual Report of the Yorkshire Union of Mechanics' Institutes* 1843: 35.

in idleness' (*Annual Report of the Huddersfield Mechanics' Institute* 1848: 13). By 1850, Huddersfield was the tenth-largest institute in Britain and the second-largest in the Yorkshire Union, with 887 members (Hudson 1851).

The Huddersfield Mechanics' Institute membership in 1857 was 662 and included 30 mechanics, two solicitors, one dentist and one architect who were obviously middle class. However, there were also 108 finishers, 60 warehouse-men and boys, 27 spinners, 24 dyers and 22 weavers from working class 'A' (*Huddersfield Mechanics' Institute Class Registers* 1857). Gray (1981) identifies that the spinners were the best-paid skilled adult workers as they supervised the work of semi-skilled piecers, thus the former would be working class 'A' and the latter working class 'B' (see Table 5.10).

Table 5.9 Known occupations of the membership at Huddersfield Mechanics' Institute 1847

Occupation	Number	Occupation	Number	Occupation	Number
Finishers	58	Clerks	15	Spinners	27
Dyers	14	Carpenters	27	Printers	12
Students	71	Mechanics	24	Errand boys, factory hands	52
Weavers	41	Warehousemen	14	Masons	13
Smiths	13	Shoemakers	14	Wool-sorters	12
Wheelwrights	12	Twiners	13	Printers	10

Source: Tylecote 1957: 455.

Table 5.10 Known occupations of the membership at Huddersfield Mechanics' Institute 1857

Occupation	Number	Occupation	Number	Occupation	Number
Finishers	108	Printers	6	Surveyors	2
Warehousemen/ boys	60	Plumbers	6	Twisters	2
Piecers	47	Smiths	6	Architect	1
Students	38	Bookbinders	5	Carder	1
Clerks	30	Coopers	5	Clogger	1
Mechanics	30	Manufacturers	5	Coal dealer	1
Spinners	27	Nippers	5	Confectioner	1
Joiners/ cabinetmakers	24	Clothiers	4	Dentist	1
Dyers	24	Feeders	4	Designer	1
Weavers	22	Slipper manufacturers	4	Druggist	1
Errand boys	19	Slubbers	4	Fish dealer	1
Masons	19	Sweeps	4	Gas worker	1
Tailors	13	Builders	3	Glass dealer	1
Grocers	12	Pupil-teachers	3	Hatter	1
Letter-carriers	12	Saddlers	3	Labourer	1
Drapers	9	Watchsmiths	3	Manager	1
Painters	8	Wheelwrights	3	Music-seller	1
Printer	8	Booksellers	2	Organ builder	1
Woollen-sorters	8	Clippers	2	Overlooker	1
Butchers	7	Colliers	2	Tea dealer	1
Millwrights	7	Engineers	2	Toy manufacturer	1
Shoemakers	7	Feeders	2	Wool-stapler	1
Carvers	6	Moulders	2		
Curriers	6	Silk-dressers	2		
Total					662

Source: *Huddersfield Mechanics' Institute Class Registers 1857* (Huddersfield: University of Huddersfield archives).

Royle (1987) has argued that warehousemen, who were the second largest group by occupation in 1857 were lower–middle class. This may have been the case previously but the class registers of the Huddersfield Mechanics' Institute for the 1850s and 1860s, which included names, occupations and addresses, confirm that the warehousemen who did attend lived in the poorer eastern district of the town and must have been on low wages (*Huddersfield Mechanics' Institution Class Registers* 1856–1868).

There were changes in names and types of occupations. At the Huddersfield Mechanics' Institute, for example, the largest group who attended in 1876 were the 483 factory operatives, making up 26 per cent of the total membership. 'Operative' was a general term for all those manual workers employed in the mills and factories (see Table 5.11 below). Warehousemen and boys made up nearly 8 per cent while mechanics only accounted for 3 per cent and engineers less than 1 per cent. Occupations include those from both working class 'A' and 'B', such as chimney sweepers, as well as teachers, mechanics and engineers, who may have been from the middling class but not necessarily.

Between 1876 and 1877 there was a fall in membership at Huddersfield due to a depression in trade. Factory operatives were still the largest group attending, down to 447 and making up 38 per cent of members, an increase of 12 per cent. Warehousemen and boys were down by only two and made up 9 per cent, an increase of 1 per cent, of membership. Membership of mechanics has also seen a slight rise, of two, and making up 5 per cent, and labourers and colliers were also up by five, contributing to 4 per cent, an increase of 1 per cent. Thus, working-class occupations continued to be well represented at Huddersfield (see Table 5.12).

Two years later, in 1879, membership had increased to 572. Of this number, 461 were factory operatives contributing 26 per cent, and warehousemen and boys made up 241 or 14 per cent. Added to this number were 138 teachers and scholars, providing 8 per cent of the membership, indicating that the Institute was offering subjects relevant to these two groups (*Forty-Second Annual Report of the Yorkshire Union of Mechanics' Institutes* 1879: 86). There were also several occupations that Neale has identified as working class 'B', such as porters and gardeners (see Table 5.13).

Finally, in 1881, eight years before the passing of the first Technical Instructions Act in 1889, which resulted in state recognition and funding for post–school education, factory operatives were still the largest group at Huddersfield with 396 or 30 per cent. Labourers and colliers made up 2 per cent, as too did the dyers, and mechanics 3 per cent. However, there were several specialist trades, which together contributed to around 74, or 6 per cent of the total membership. Also, manufacturers, chemists and schoolmasters were now prominent, indicating that the Institute had something to offer, in relation to subjects, across the social divide (see Table 5.14).

At the Stockton Mechanics' Institute, for example, in 1849, there had been a substantial increase in working-class membership of up of 200 journeymen and assistants and 100 apprentices, making up 75 per cent of its total membership.

Table 5.11 Known occupations of the membership at Huddersfield Mechanics'
Institute 1876

Occupation	Number	Occupation	Number	Occupation	Number
Factory operatives	483	Engineers	13	Wiredrawers	4
Bookkeepers	158	Letter-carriers	12	Sawyers	4
Warehousemen and boys	113	Marble masons	11	Basket-makers	3
Errand boys	92	Moulders	11	Brokers	3
Joiners, cabinetmakers	83	Rug-makers	11	Saddlers	4
Schoolboys	75	Plasterers	10	Ministers	4
Mechanics	56	Designers	10	Cloggers	3
Teachers	53	Porters	10	Soda-water makers	3
Labourers, colliers	38	Ironmongers	10	Surgeons	3
Masons	37	Pattern-makers	10	Chimney sweepers	3
Dyers	31	Wheelwrights	9	Hawkers	3
Draper's assistants	30	Timekeepers	8	Pianoforte makers	2
Woollen manufacturers	27	Watchmakers	8	Drysalters	2
Printers	27	Bookbinders	8	Colour-makers	2
Smiths	26	Wool-sorters	8	Engravers	2
Greengrocers	26	Organ builders	8	India rubber manufacturers	2
Shop assistants	22	Rope-makers	7	Page boys	2
Painters	19	Nippers	6	Confectioners	2
Chemists	17	Butchers	6	Carvers, gilders	2
Leather curriers	16	Surveyors	6	Bakers	2
Railway servants	16	Brickmakers	6	Clothes dealers	2
Messengers	16	Lithographers	6	Hairdressers	2
Tobacco twisters	16	Card-makers	6	Compositors	2
Plumbers	16	Upholsterers	5	Coopers	2
Farmers, gardeners	16	Brassfounders	4	Waste dealers	1
Wool-turners	14	Stokers	4		
Tailors	13	Tinners	4		
Total					1537

Source: *Thirty-Eighth Annual Report of the Yorkshire Union of Mechanics' Institutes* 1876: 160.

When the Institute had first opened in 1825 there had been only two journeymen
(*Eleventh Annual Report of the Yorkshire Union of Mechanics' Institutes* 1849:
86). The Darlington Mechanics' Institute Committee reported a change in membership in 1859 stating that:

> although Mechanics' Institutions have not been so generally supported by the
> class for whose benefit they were originally intended, the Committee is glad to
> say the Darlington Institute includes a fair proportion of working men amongst

Table 5.12 Known occupations of the membership at Huddersfield Mechanics' Institute 1877

Occupation	Number	Occupation	Number	Occupation	Number
Factory operatives	447	Joiners, cabinetmakers	76	Labourers, colliers	43
Bookkeepers and clerks	158	Schoolboys	74	Masons	35
Warehousemen and boys	111	Mechanics	58	Shop assistants	33
Errand boys	89	Teachers	56		
Total					1180

Source: *Thirty-Ninth Annual Report of the Yorkshire Union of Mechanics' Institutes* 1877: 129.

Table 5.13 Known occupations of the membership at Huddersfield Mechanics' Institute 1879

Occupation	Number	Occupation	Number	Occupation	Number
Factory operatives	461	Building trades	220	Warehousemen and boys	241
Teachers, scholars	138	Iron trades	158	Woollen manufacturers	62
Leather trades	21	Tailors	8	Bookkeepers, clerks	158
Printing, bookbinding	39	Porters, letter-carriers	31	Solicitors, accountants	39
Shopkeepers	47	Farmers, gardeners	11	Watchmakers	9
Cabinetmaking	21	Designers	17	Dyers	25
Miscellaneous	46				
Total					1752

Source: *Forty-First Annual Report of the Yorkshire Union of Mechanics' Institutes* 1879: 111.

its members. Such associations open to every class and presenting a common ground of union, commend themselves to the support of all interested in the scope of the labours and generous rivalry of this kindred Institution (*Twenty-First Annual Report of the Yorkshire Union of Mechanics' Institutes* 1859: 76).

Neale and Hobsbawm complement and support the argument here that occupations of members who attended the mechanics' institutes after 1830 were working class. Hobsbawm (1964) in particular is more specific about what occupations he sees as being of the labour aristocracy and actually increases the number of occupations of those institutes discussed here as being working class. Thus, workers, often from the more skilled trades, together with those from the lower middle classes, were also the ones who aspired to intellectual development, previously reserved for the 'higher reaches of society' (Royle 1987: 248). Mechanics' institutes were ideally suited to supporting these classes.

Table 5.14 Known occupations of the membership at Huddersfield Mechanics' Institute 1881

Occupation	Number	Occupation	Number	Occupation	Number
Factory operatives	396	Pattern-makers	6	Organ builders	2
Clerks, office boys	137	Tinners	6	Corn dealers	2
Warehousemen and boys	65	Woodcarvers	6	Engravers	2
Errand boys	64	Grooms	6	Foremen	2
Joiners, cabinetmakers	62	Gas fitters	5	Packing maker	1
Schoolboys	56	Whitesmiths	5	Porter	1
Pupil-teachers	45	Marble masons	5	Apprentice	1
Mechanics	43	Moulders	5	Clipper	1
Masons	36	Travellers	5	Artist	1
Manufacturers	34	Coachbuilders	4	Soda-water maker	1
Labourers, colliers	29	Brick moulders	4	Pork butcher	1
Dyers	29	Hawkers	4	Milk-seller	1
Grocers	25	Waste dealers	4	Card-nailer	1
Shop boys	21	Rope-makers	4	Clothier	1
Printers	20	Plasterers	4	Dentist	1
Engineers	20	Butchers	4	Billiard proprietor	1
Painters	18	Photographers	4	School Board Visitor	1
Bookkeepers	15	Drysalters	4	Cooper	1
Chemists	14	Architects	4	Wine merchant	1
Schoolmasters	14	Merchants	4	Veterinary surgeon	1
Designers	11	Nippers	3	Goldsmith	1
Blacksmiths	10	Bookbinders	3	India rubber maker	1
Drapers	10	Finedrawers	3	Waiter	1
Gardeners, farmers	9	Firemen	3	Agent	1
Wheelwrights	9	Upholsterers	3	Wire drawer	1
Jewellers	8	Cloggers	3	Stoker	1
Rug-makers	8	Pipemakers	3	Oil merchant	1
Telegraph boys	8	Accountants	3	Police officer	1
Ironmongers	8	Tobacconists	3	Brushmaker	1
Brass finishers	7	Letter-carriers	3	Dressmaker	1
Confectioners	7	Timekeepers	3	Miller	1
Tailors	7	Potters	2	Salter	1
Shoemakers	7	Curriers	2	Carrier	1
Surveyors	7	Wool-sorters	2		
Plumbers	6	Bottle-washers	2		
Hairdressers	6	Sawyers	2		
Total					1443

Source: *Forty-Third Annual Report of the Yorkshire Union of Mechanics' Institutes* 1881: 101.

Neale's five-class model provides the appropriate categorisation for identifying class in relation to occupations of members of mechanics' institutes and which supports the main hypothesis of this study, that the majority of members of institutes after their early years were working class and specifically from working class 'A'.

The evidence provided in relation to occupations of members of several Yorkshire Union institutes which has survived through the Annual Reports of the Yorkshire Union, indicates that there was a change in the occupational membership of many mechanics' institutes, from middle class towards more working class, as defined by Hobsbawm and Neale. Smaller institutes did not include a listing of members' occupations, in some cases only sending in brief reports, but with the evidence provided above and the known locations of the smaller institutes, in rural areas and small textile communities, the evidence provided puts beyond doubt that there was a large working class membership.

Further evidence of working-class occupational membership in Mechanics' Institutes

Another very important indicator that the mechanics' institute membership was becoming more working-class, was the changing attitudes of many mechanics' institute committees. At Middlesbrough, for example, as early as in 1849, the Committee was conscious that the welfare of the Institute depended on the activities and energies of working-class members, 'who could understand and appreciate the wants of the working classes' (*Twelfth Annual Report of the Yorkshire Union of Mechanics' Institutes* 1849: 71).

The Hartlepool Institute in 1860 identified the need, through putting out an appeal, for working-class men to become Committee members in order that they had ownership of decision-making (*Twenty-Second Annual Report of the Yorkshire Union of Mechanics' Institutes* 1860: 86). The Committee at Stockton reported in 1861 that:

> it was gratifying ... to state that the proportion of working men belonging [to] the Institute is on the increase. Indeed, out of the 348 members, 186 [58 per cent] consisted almost exclusively of working men and working apprentices, more than one half [of] the entire number of members (*Twenty-Third Annual Report of the Yorkshire Union of Mechanics' Institutes* 1861: 128).

The Committee was enthusiastic about encouraging the working class to attend the Institute, sending out a plea to employers that it respectfully urged

> upon the firms, employing large numbers of workmen, and the members generally, to use efforts to increase the number of members, so as to enable them to spend still more on the classes, which they [the Committee] believe to be the most important means of usefulness in Mechanics' Institutes

(*Twenty-Fourth Annual Report of the Yorkshire Union of Mechanics' Institutes* 1862: 130).

Other mechanics' institutes experiencing working-class involvement included the Keighley Institute, which reported in 1838 its pride in the fact that it was established by four working-class men, a reed-maker, a painter, a tailor and a joiner, who 'were anxious to secure for their own town and neighbourhood the benefits to be derived from such societies as had recently been established in Glasgow, Edinburgh, London, Liverpool, and a few other places' (*Twenty-First Annual Report of the Yorkshire Union of Mechanics' Institutes* 1859: 95). When Huddersfield reopened in 1841, it was established by a small group of employees of Frederic Schwann, who had originally provided a library for his employees (*Fifth Annual Report of the Yorkshire Union of Mechanics' Institutes* 1843: 25).

The Committee reported in 1853 to the Yorkshire Union that it was

> mainly working men – men engaged all day long in the mills, factories, and workshops – who have come night after night for years together, and without any other reward except the approbation of their own consciences, have devoted themselves to its service, either as committee-men, teachers, or librarians,

as well as students (*Fifteenth Annual Report of the Yorkshire Union of Mechanics' Institutes* 1853: 74). Thus, not only were the students working class, but so were many members of staff, a development in several institutes in the district encouraged by committees, adding to the sense of a real pride in themselves as working-class self-educators who knew first-hand the educational needs of the working-class membership.

In the Yorkshire Union, there is substantial evidence of the establishment of working-class mechanics' institutes. While the larger general mechanics' institutes founded in the 1820s and 1830s, such as those at Keighley and Skipton in the Yorkshire Dales and at Darlington, Hartlepool, Middlesbrough and Stockton in the North East, those established in the smaller communities such as in the agricultural districts of the Dales and Pennines, and mining communities of the North East, had predominately working-class membership from their foundation. It was not just mechanics' institute committees that reported working-class involvement. Barnet Blake, a Yorkshire Union agent and lecturer, observed that in the case of the Yorkshire Union at least, the majority of institutes not only provided 'educational wants of working men' but they were also 'mainly supported, and, in many instances, managed by them' (Blake 1859: 335).

Conclusion

Contemporaries and historians have argued that these mechanics' institutes established by the middle class were not the success they had hoped they would be in providing working-class education and that, by 1850, they were in decline.

Historians have therefore assumed that the movement 'failed' in its objective, especially as many studies related to the movement up to 1850 reflect this. The majority of historians that have written on mechanics' institutes are convinced that they were established, managed, and patronised by the middle class. Several argue that the early founders, such as Birkbeck and Brougham, reinforced middle-class ideology. The public lectures and teaching were of a scientific and intellectual level that discouraged the working classes from attending in any case, all of which contributed to their failure.

However, over several years, as the membership became much broader, mechanics' institutes were also being patronised by the working class, as evidenced by the work of Thompson, Hobsbawm and Neale. Their work, relating to several specific examples of occupations of members at Yorkshire Union institutes, confirms that there was an increase in the number of working-class members attending them. The 'old' skilled occupations were being replaced with industrial ones associated with the labour aristocracy. There were also members with occupations associated with both working class 'A' and working class 'B' as identified by Neale. Working class 'B' included miners who had the opportunity to attend mechanics' institutes being established in the new developing coal and iron stone mining communities of the North East and rural workers in the agricultural and small-scale industries in the Dales and Pennines.

The membership was not only made up of a growing number of working-class attendees at mechanics' institutes, but there is also substantial evidence that many institutes encouraged them to be part of the committees. Such an opportunity not only encouraged new working-class members but also enabled them to inform their institutes of what subjects and support were required to continually attract new members. The Committee at Skinningrove Mechanics' Institutes put it succinctly in 1882, when it stated that 'its march of usefulness is standing proof that mechanics' institutes can be made conversant to the welfare of the working population (*Annual Report of the Yorkshire Union of Mechanics' Institutes* 1882: 147).

By 1900, the East Lancashire textile town of Burnley was supporting its working-class population:

> the mill, mine, and engineering workers of Burnley were known as hard grafters, working long hours for high rewards. Many of the workforce gave up their spare time to enrol in evening educational classes at the Mechanics' Institute to gain further skills and qualifications that would make them higher paid and more employable than in other towns (Kelly 1966: 238).

Referring back to the opening paragraph, Lady Bracknell had nothing to worry about with regard to the education of the working classes as, although there were opportunities for some of them at least, they did not prove a serious danger to the upper classes or lead to acts of violence in Grosvenor Square or indeed anywhere else. Whilst coming from the polite London society, she had no inkling as to what was happening in parts of England where mechanics' institutes had an active working-class membership.

References

Blake, B. (1859) 'The Mechanics' Institutes of Yorkshire', *Transactions of the National Association for the Promotion of Social Sciences 1859*, 335–340.

Briggs, A. (1960) 'The language of "class" in early nineteenth-century England', Briggs, A., and Saville, J. (eds) *Essays in Labour History* (London: Macmillan): 43–73.

Elliot, R. (1861) 'On the working men's reading rooms, as established in 1848 at Carlisle', *Transactions of the National Association for the Promotion of Social Science 1861*, 673–677.

Gaskell, E. (1967) *Mary Barton, A Tale of Manchester Life* (Reprint. Oxford: Oxford University Press).

Godwin, J. V. (1860) 'The Bradford Mechanics' Institute', *Transactions of the National Association for the Promotion of Social Science 1859*, 315–340.

Gray, R. (1981) *The Aristocracy of Labour in Nineteenth-Century Britain c.1850–1914* (London: Macmillan).

Harris, J. (1993) *Private Lives, Public Spirit: Britain 1870–1914* (London: Penguin).

Hobsbawm, E. J. (1964) *Labouring Men, Studies in the History of Labour* (London: Weidenfeld & Nicolson).

Hobsbawm, E. J. (1984) *Worlds of Labour, Further Studies in the History of Labour* (London: Weidenfeld & Nicolson).

Huddersfield Mechanics' Institute (1841) *Annual Report*.

Huddersfield Mechanics' Institute (1848) *Annual Report*.

Huddersfield Mechanics' Institute (1857) *Class Registers* (Huddersfield: University of Huddersfield archives).

Huddersfield Mechanics' Institute (1856–1868) *Class Registers* (Huddersfield: University of Huddersfield archives).

Huddersfield Scientific and Mechanic Institute (1825) *Rules of the Huddersfield Scientific and Mechanic Institute, for the Promotion of Useful Knowledge*, (Huddersfield).

Hudson, J. W. (1851) *The History of Adult Education in which is comprised a Full and Complete History of the Mechanics' and Literacy Institutions* (Reprint. London: Woburn, 1969).

Hudson, P. (1992) *The Industrial Revolution* (London: Bloomsbury).

Inkster, I. (1975) 'Science and the Mechanics' Institutes, 1820–1850: The case of Sheffield', *Annals of Science* Vol. 32, No. 5, 451–474.

Inkster, I. (ed.) (1985) *The Steam Intellect Societies, Essays on Culture, Education and Industry circa 1820–1914* (Nottingham: Nottingham University).

Kelly, T. (1966) *Early Public Libraries: A History of Public Libraries in Great Britain before 1830* (London: The Library Association).

Langley, J. B. (1849) *Evidence to the Select Committee on Public Libraries (1849)* Appendix III, Royle, E. (ed.) (1971) 'Mechanics' Institutes and the working classes, 1840–1860', *Historical Journal* Vol. XIV, No. 2, 305–321.

Luckhurst, K. W. (1957) *Some Aspects of the History of the Society of Arts*, unpublished PhD thesis (London: University College London), Chapter X, 4.

McWilliam, R. (1998) *Popular Politics in Nineteenth-Century England* (London: Routledge).

Neale, R. S. (1981) *Class in English History 1680–1850* (Oxford: Blackwell).

Perkin, H. (1986) *Origins of Modern English Society* (London: Ark Paperbacks).

Royle, E. (1971) 'Mechanics' Institutes and the working classes, 1840–1860', *The Historical Journal* Vol. XIV, 2, 305–321.

Royle, E. (1987) *Modern Britain, A Social History 1750–1985* (London: Arnold).

Select Committee on Education (1853) *Report from the Select Committee on Education, Manchester and Salford, and Proceedings, Minutes of Evidence* (London: House of Commons).

Stockdale, C. (1994) *Mechanics' Institutes in Northumberland and Durham 1824–1902*, PhD thesis (Durham: Durham University).

Thompson, E. P. (1968) *The Making of the English Working Class* (London: Penguin).

Tosh, J. (1999) *A Man's Place: Masculinity and the Middle-Class Home in Victorian England* (Bath: Bath Press).

Tylecote, M. (1957) *The Mechanics' Institutes of Lancashire and Yorkshire Before 1851* (Manchester: Manchester University Press).

West Riding of Yorkshire Union of Mechanics' Institutes (1838) *First Annual Report.*

Wilde, O. (1998) *The Importance of Being Earnest and other Plays* (Oxford: Oxford World's Classics), 253–307.

Yorkshire Union of Mechanics' Institutes (1840) *Second Annual Report.*

Yorkshire Union of Mechanics' Institutes (1841) *Third Annual Report.*

Yorkshire Union of Mechanics' Institutes (1843) *Fifth Annual Report.*

Yorkshire Union of Mechanics' Institutes (1844) *Sixth Annual Report.*

Yorkshire Union of Mechanics' Institutes (1849) *Eleventh Annual Report.*

Yorkshire Union of Mechanics' Institutes (1853) *Fifteenth Annual Report.*

Yorkshire Union of Mechanics' Institutes (1859) *Twenty-First Annual Report.*

Yorkshire Union of Mechanics' Institutes (1860) *Twenty-Second Annual Report.*

Yorkshire Union of Mechanics' Institutes (1861) *Twenty-Third Annual Report.*

Yorkshire Union of Mechanics' Institutes (1862) *Twenty-Fourth Annual Report.*

Yorkshire Union of Mechanics' Institutes (1876) *Thirty-Eighth Annual Report.*

Yorkshire Union of Mechanics' Institutes (1877) *Thirty-Ninth Annual Report.*

Yorkshire Union of Mechanics' Institutes (1879) *Forty-First Annual Report.*

Yorkshire Union of Mechanics' Institutes (1881) *Forty-Third Annual Report.*

Yorkshire Union of Mechanics' Institutes (1882) *Forty-Fourth Annual Report.*

6 Mechanics' institutes and female membership

Introduction

Schooling in the 1850s reinforced gender differences. Boys were more likely to attend school, from the ages of six or seven, while girls usually worked and were trained at home by their mothers. Those girls that did go to school tended to learn domestic subjects alongside reading, writing and arithmetic. Children of very poor families did not attend school and were therefore expected to help around the home from an early age with little differentiation between girls and boys and with no opportunities for education either as children or, later, adults. This was to continue until the passing of the 1870 Education Act, which was the beginning of state support for all. Having said that, primary education for the poor had been available, for some at least, through Dame and Charity schools (Shoemaker 1998: 131).

Shoemaker believes that the Chartist movement, while it allowed wives and daughters of its supporters to set up and run Chartist schools, in all other ways excluded them from taking part in what was seen as male radical politics. Women 'therefore justified their participation in Chartism by emphasising their moral duty to support the welfare of their families and neighbours, supporting the men but not attempting to supersede or rival them' (Shoemaker 1998: 263). He argues that Chartist men embraced working-class women having middle-class values such as education, provided they related to domestic respectability. By arguing that women, even among the lower classes, should not work outside the home and should devote themselves to domestic duties, while men were the breadwinners, Radicals defused tensions created by competition between men and women for jobs and re-established the traditional pre–industrialised balance of power within the family economy. Thus, working-class education, where it existed, was male-dominated, with females expected to learn domestic skills at home.

Rose (1992: 22–3) defines gender as depicting differing positions of women and men in nineteenth-century society. In practice, positions of women and men were different and that structured their experiences in different ways. She highlights that, during the nineteenth century, nearly all employers hired men for skilled work involving 'complicated machinery'. Indeed, '[employers] rarely questioned the appropriateness of hiring male workers for jobs' as being responsible for running large and complex machines was seen as a masculine trait, and manufacturers hired women for work which they thought was 'women's work'. However, Rose identifies that the situation in relation to gender and employment was somewhat different with regard to textile industries. In Lancashire, for example, power loom weaving 'was unique among the major industries of England in the nineteenth century in that men and women were not in competition for weaving jobs' (ibid.: 154). She points out that men and women were paid the same piece rates and this was supported by the weavers' unions which represented females as well as males.

Hudson (1992: 230) points out that with industrialisation there was a greater involvement of women in waged work outside the home and this 'brought greater parity between partners'. She also highlights that female Chartists were predominant in the Temperance Movement, which was a supporter of the Mechanics' Institute Movement. Hudson also observes that working-class women were not subjected to the same scrutiny as their counterpart upper classes and had the 'freedom' to work alongside men in the weaving mills and coal-mining districts, both of which were in abundance across the North and Midlands.

Davidoff and Hall (1998: 262) rightly claim, however, that women from the:

> middle and lower ranks were culturally most disadvantaged. Their sole avenue of mental cultivation was the Sunday school. The mechanics' institutes and literary and philosophical societies which were so important for their male equivalents were usually closed to young women.

In any case, Royle (1987) believes that, while literacy levels were improving among some adult working-class male groups, women generally lagged behind, even if they had been fortunate enough to have had an elementary school education. Female employment was likely to have involved little education apart from some reading. As most dealings outside the home were carried out by men, any education gained by women was often forgotten as it was not practiced. Men also had the opportunity to attend outside organisations such as clubs, unions and political associations. Royle also states that men did have the opportunity to attend mechanics' institutes and receive basic instruction 'in the elements of reading, writing and arithmetic'. He cites Birmingham Mechanics' Institute, which had a membership in the late 1830s of about 485 members, none of whom were women (Royle 1987: 352). It is hardly surprising that when mechanics' institutes began to respond to the needs of adult working-class education it was very much assumed only men would be interested in attending.

It was only after 1850 that working-class men were becoming engaged with their local institutes relevant to work-related subjects. Like them, working-class women had had little or no schooling during the 1830s and 1840s. However, from 1849 the Yorkshire Union of Mechanics' Institutes began to receive the number of females attending several institutes, albeit on much smaller numbers than men.

However, as discussed previously, certainly philosophical societies were more for the professional classes interested in attending scientific lectures rather than elementary and technical education offered by the mechanics' institutes.

Watts (1998: 153) states that although the science and literary societies were promoted by Unitarians, who were supportive of female rights, women were invited as guests to attend selected events; 'at best they were only second-class members'. Purvis (1991: 37) confirms that the aim of the Mechanics' Institute Movement was 'both class and sex specific that is, the diffusion of scientifically useful knowledge to working-class men'. This is a fair observation of course up until the 1840s. However, Purvis has calculated that in England and Wales by 1851 there were 5,710 women and 55,239 men who were members of their local institute. Of the 5710 women, at least 1200 were attending Yorkshire institutes and by 1860 this had increased to at least 2737, and, as not all institutes in Yorkshire were members of the Union, this may well be a low estimate (*Twelfth Annual Report for the Yorkshire Union of Mechanics' Institutes* 1850 and *Twenty-Second Annual Report for the Yorkshire Union of Mechanics' Institutes* 1860, statistical returns). Purvis (1991: 42) identified that some mechanics' institutes in Yorkshire, such as those at Pudsey and Halifax in the West Riding, did take female members who tended to study the three 'R's. However, she did not extend her research beyond the 1860s. Her work is backed up by Brown and Inkster (2000) who calculated from the 1851 Census that literary and scientific institutions generally had the highest number of female members in Kentish London, parts of the Midlands and the West Riding of Yorkshire. Devon, Essex and Cornwall also had active female membership.

Harris (1993: 115) identifies that working-class women were treated as subordinates to their male counterparts who 'treated them as weaker workers after the 1870s'. She highlights that 'a small minority of working women, mostly weavers, got equal pay and status with men' but for the majority of females they were viewed by their male counterparts as rivals or subordinates rather than fellow members of the same class and workforce. Even among working class women themselves there was differentiation. There was a 'great divide between the "hat girls" and "shawl girls" in the Lancashire mill towns' and in London, factory girls, costermongers (sellers of fruit and vegetables from a barrow or stall in the street) and domestic servants viewed each other 'with as much contempt as a belle of New York and a Chicago heiress' (ibid.: 146).

Watts states that from 1850 there were revolutionary changes in female education. She identifies various causes, economic, demographic, social, psychological and religious, as well as individual influences particularly from Unitarians, one of a number of groups that supported the Mechanics' Institute Movement. Watts rightly argues that women were as intellectual as men, although there were physiological differences. The Unitarian movement was ideally positioned from 1850 onwards to support a 'full education for women and greater rights and opportunities for them' through mechanics' institutes (Watts 1980: 286). Harriet Martineau, writing in 1851, stated that mechanics' institutes were enthusiastic in support of women's education, in contrast to the universities 'where intellectual luxury was reserved to pamper the few while the many starved' of education (Martineau 1851: 111). Some institutes were only able to offer a minimum number of courses, and usually at elementary level, but the 'two chief single-sex female institutes did provide a wider curriculum (Watts 1998: 186–7). These were located in Bradford and Huddersfield. Three of the most popular institutes for females, as well as men, were Manchester, Liverpool and Leeds. The first two were under Unitarian management, and in the case of Liverpool they supported both boys' and girls' schools which were attached to the Institute (ibid.: 188).

Many institutes seem to have opened their doors to females in regard to public lectures and were certainly patronised by middle-class women initially, among them Elizabeth Gaskell who attended lectures at the Manchester Institute (ibid.: 189). The Bronte sisters attended Keighley Institute where they attended the lectures and made use of the library. The Rev. Patrick Bronte became a member of the Institute in 1833 and his children had access to both the library and scientific lectures. The Bronte family attended the Institute between 1835 and 1855 and Charlotte Bronte gave talks at both Keighley and Haworth Institutes. A 'programme of fortnightly lectures, which included demonstrations of chemistry and talks on subjects as diverse as physical astronomy and the wheel and axle' were available at Keighley for both men and women (Barker 2010: 171). Elizabeth Gaskell, herself a great supporter of mechanics' institutes, met Charlotte Bronte in 1850 and they became close friends. After Charlotte's death, Gaskell wrote her biography (Barker 2010).

The mechanics' institutes were ideally suitable as educational institutions to accept women as both students and teachers for the female classes, in domestic and elementary education (O'Day 2000). Skedd (1997) has identified that it was middle-class women who trained as school teachers and had been educated at boarding schools. Some supported the mechanics' institutes but in many cases it was former students who were predominantly working-class women like their male counterparts, who became involved in teaching the female classes.

Institutes identified that they should encourage women to become members, which seems to have taken place with little opposition. However, there are examples where the decision to include women was not unanimous, despite the potential for increasing the membership and income. George Searle Phillips, a journalist and writer who became Secretary of the Huddersfield Mechanics' Institute between 1846 and 1854 and agent for the Yorkshire Union between 1846 and 1856, for example, argued that women should be educated as housewives, not scholars:

> A 'bluestocking' – that is, a female literary pedant – is certainly no desirable person to know either in private or public; but there is no necessity to manufacture this kind of hosiery on our educational looms … I do not see why a woman should not be a good housewife, and as prudent, virtuous, and honourable in all her relations with a furnished as with an unfurnished mind (Purvis 1989: 101).

Purvis (1989) found that the number of women in the institutes was highest in the Yorkshire and Lancashire and Cheshire Unions of Mechanics' Institutes. Dr James Kay-Shuttleworth, who wrote *The Moral and Physical Condition of the*

Working Class Employed in the Cotton Manufacture in Manchester (1832), which was cited by Friedrich Engels in *Condition of the Working Class in England in 1844* as well as papers on both medicine and education, was President of both the East Lancashire Union of Mechanics' Institutes and the Burnley Mechanics' Institute, and introduced subjects and examinations for women from 1858. Examinations were held in domestic hygiene and economics, and management of children and sick nursing (Shacklady [no date]: 2).

Tylecote (1957) estimated that there were about 1200 women members in the Yorkshire Institutes in 1849, and about 600 in Lancashire. The higher number of women in the northern counties may be partly explained by the fact that the Yorkshire Union during the 1840s consciously supported the cause of female education despite the views held by Phillips at Huddersfield. There were also similar employment opportunities for males and females in the textile and mining districts of the North as previously discussed. Although women formed only a small proportion of the national membership – about 9 per cent – the proportion of female members varied between 3.5 per cent and 22.4 per cent, the large institutes recruiting the most. Yet the figure of 5710 women was not insignificant and Purvis (1989: 109) believes that these women were working class.

Female membership in the Yorkshire Union of Mechanics' Institutes

The debates around female education and the mechanics' institutes will be taken up in relation to those institutes that were members of the Yorkshire Union which are known to have sent in female membership returns with their annual reports. Evidence suggests that women were a more important part of the movement than previously suggested. If this is the case, then the Mechanics' Institutes Movement could well have been the forerunner for providing female education which previously had been assumed to be at best available for middle-class girls through finishing schools, and at worst no education for working-class girls and females which provided no formal education opportunities.

An article entitled 'Practical Female Education' was published in the 1858 *Annual Report of the Yorkshire Union of Mechanics' Institutes*. The unknown author made a case for admitting females into mechanics' institutes, the argument being that 'improvidence is the great bane of the working classes. They live from hand to mouth, have no desire to save, and do not possess the all-important art of making a little go a long way' (*Twentieth Annual Report of the Yorkshire Union of Mechanics' Institutes* 1858: 34–5). The argument put forward is that working-class families in the industrial towns earned an income that was higher than that in the countryside. However, any reduction in working hours meant a substantial fall in income and any stoppages in the mills, or ill health, resulted in abject poverty. Thus, to teach both women and girls principles in domestic economy, cooking and dressmaking would, so it was believed, support financial security in the home. The article believed mechanics' institutes should offer such courses to adult females and that National Schools should provide them for girls. It was on this basis of home economics and dressmaking that female membership evolved in the Yorkshire Union.

In 1857 the Yorkshire Union advised those wishing to establish an institute in a small town or village that they should include female classes for reading, writing and needlework (*Nineteenth Annual Report of the Yorkshire Union of Mechanics' Institutes* 1857: 25). In 1859 Barnett Blake, agent and lecturer of the Yorkshire Union, stated that such a curriculum was 'so necessary' to the comfort of the working-class household.

> The great importance of female education is one of the surest means of achieving social amelioration. As the first lessons of instruction, whether good or evil, are derived from the mother, it is evident that our young females should not be neglected. Upon their training depends much of the future (ibid.: 10).

Table 6.1 lists the subjects offered to women at selected institutes between 1846 and 1861. As with male classes, the subjects common in all co-educational institutes were those offered at elementary level. It is interesting to note that at Holmfirth, branches of science were also offered to females.

Table 6.1 Selected Institutes and subjects offered to female members

Institution	Curriculum offered to women	Year
Holmfirth	Access to the library and lectures. Classes offered in reading, writing, arithmetic, sewing and knitting, general knowledge, geology, electricity, galvanism, properties of water and phrenology.	1846
Kirkstall	Female teacher read to the students in the sewing class, 'giving a familiar language such explanations as they need; thus instructing and amusing their minds while skill with a needle was acquired' (1846: 39).	1846
Wakefield	Reading, writing and arithmetic.	1847
Holbeck	Reading, writing and arithmetic.	1847
Wilsden	Classes for women were 'for the improvement of their minds' (1848: 86).	1848
Pudsey	As well as the three 'R's, women could study dictation, drawing and phonology, the branch of science that dealt with speech sounds and pronunciation that is now known as linguistics and phonetics.	1852
Halifax	Reading, writing and arithmetic, sewing, dictation, grammar, geography and history.	1861

Sources: *Eighth Annual Report of the Yorkshire Union of Mechanics' Institutes* 1846: 36–39; *Ninth Annual Report of the Yorkshire Union of Mechanics' Institutes* 1847: 72–92; *Tenth Annual Report of the Yorkshire Union of Mechanics' Institutes* 1848: 86; *Fourteenth Annual Report of the Yorkshire Union of Mechanics' Institutes* 1852: 76; *Twenty-Third Annual Report of the Yorkshire Union of Mechanics' Institutes* 1861: 78.

Fanny Hertz of the Bradford Female Institution published a paper in 1860, entitled 'Mechanics' Institutes for working women, with special reference to the manufacturing districts of Yorkshire'. She stated that, while the education of working women was crucial for both social happiness and progress, it had not been given the high priority it deserved:

> During the last third of the century working men have been initiated, and have rapidly advanced in nearly every department of knowledge, while their minds are ever expanding more and more under the beneficent stimulus they constantly receive. Women, on the other hand, were not so fortunate in having these opportunities generally, of awakening in their fellow female companions a corresponding desire after a higher moral and intellectual life' (Hertz 1860: 347).

Hertz argued that while the education of working men was advantageous in their employment and life, generally the same was not true for women. She did reinforce that it was the Unitarian view of female education that was acceptable, namely in support of 'the duties of wives and mothers, of mistresses and servants'. Hertz accepted that as social and political opportunities for working men developed, so they would too for women.

Hertz used as her research case studies the female institutions of Bradford and Huddersfield. In the case of Bradford, she was frustrated at seeing the state which working-class women were in as a result of social and financial deprivation, particularly poor diet and appearance. Hertz put this down to little or no education, which, if available, would help improve living conditions of working-class families and support women with the knowledge for social reform to support both working-class men and women.

In the case of Bradford, the working-class females had a lot in common with their male counterparts:

> the female population, from its employments and its habits of life, has much in common with male artisans. The mechanics' institutes, which have effected such an incalculable amount of good in the case of working men, will no doubt also prove to be the agencies best fitted to awaken in the analogous classes of the other sex those higher aspirations and faculties now dormant, and to give them the first impulse to self-culture (ibid.: 350).

In 1860, the Bradford Female Education Institute had as its main objective to 'provide for females of this town and neighbourhood increased faculties for mental improvement by means of classes, a Library, addresses and other such methods as may appear suitable for imparting sound moral and secular instruction' (Adams 1970: 202). The Committee at the Bradford Female Institute in 1862 provided a list of the trades associated with its female members. Table 6.2 highlights that the occupations were working class and that the courses must have been relevant to the industries women were working in. For two d. (two

Table 6.2 Known occupations of members at the Bradford Female Education Institute 1862

Occupation	Number	Occupation	Number
Weavers and twisters	129	Spinners, other mill workers	267
Nurse-maids at home	122	Dressmakers	6
Assistant teachers	2	Piece-board paperers	10
Domestic servants	17	Cigar-maker	1
Employed in chemical works	1	Employed in paper warehouse	3
Employed in shops	4	Stay-makers	2
Upholsterers	2	Bookbinders	2
Stitchers in dye-house	2		
Total			570

Source: *Twenty-Fourth Annual Report of the Yorkshire Union of Mechanics' Institutes* 1862: 83.

pre-decimal pence) a week 'young women and mothers whose education has been neglected in their early days' could study sewing, cookery and general subjects four evenings a week.

Plain sewing and dressmaking classes were seen as important economically. Once they had made the clothes, on payment for the raw materials, they could take them home. The Committee held quarterly meetings for the purpose of managing the Institute with teachers attending the regular meetings supporting curriculum and teaching developments (*Twenty-Seventh Report of the Yorkshire Union of Mechanics' Institutes* 1865: 89).

In her observations, Hertz noted that Huddersfield was 'a far prettier and more cheerful town [than Bradford] with hardly a street from which green fields may not be seen'. Women were 'not engaged [to] so large an extent in factory work, nor are they of that unsettled wandering disposition which so often baffles our efforts to educate them in Bradford'. Her observations were correct as the female students who attended the Institution in Huddersfield tended to be milliners, dressmakers, domestic servants or daughters still living with their parents. They did not seem as independent as their Bradford counterparts, and Hertz noted that 'there exists between them and the managers of their Institute a warm friendliness, and personal regard, productive of the best results' (1860: 351).

The Huddersfield Female Institution was founded in 1847 and the Bradford Female Educational Institution was opened in 1859, less than two years before Hertz wrote the article. Although there were 600 women enrolled at Bradford, which was a larger town than Huddersfield, the membership fluctuated and Hertz believed that by improving the teaching, the numbers would become more stable or expand and the Institute would be a success (ibid.).

Subjects offered at Bradford were reading, writing, arithmetic, geography, history, grammar, needlework and singing, as well as 'an advanced class for the elements of natural science (human physiology), which was seen as important for potential mothers)' (ibid.: 353). There was a library, reading room and penny savings bank, and public lectures were delivered from time to time. Hertz believed that 'those who assist to establish mechanics' institutes for young women, or who in any way further the mental improvement of female operatives, will be providing suitable and worthy helpmates for the educated and intelligent working man' (ibid.: 354).

The Female Education Institute was established in Huddersfield by Frederick Schwann and supported and organised by his wife, the daughter of a Unitarian minister from Birmingham. It seems this was a personal choice, to educate females separately from males in different institutes. The Female Educational Institute, opened in 1847, reported in 1857 that it had been originally founded 'by Ladies and Gentlemen [who] observed in many villages, and most towns of the Kingdom, mechanics' institutes for the instruction of young men, but no similar provision for the young women of the working classes' (*Nineteenth Annual Report of the Yorkshire Union of Mechanics' Institutes* 1857: 85). Robert Baker, a factory inspector for Leeds, noted that there were 409,360 females at work 'within the walls of the factories in the United Kingdom', and believed strongly that similar institutions to the Female Institute at Huddersfield were needed to provide educational opportunities for women, as well as men (*Huddersfield Female Institute Minutes of the General Committee* 1862: 8). The *Eighth Annual Report of the Yorkshire Union of Mechanics' Institutes* in 1846 had raised questions in relation to the 'state of education amongst young women of the same [working] class' as men. The following year, the annual report for the male Institute stated that 'the attentions of your Committee has been called during the past year to the propriety of endeavouring to afford to the female portion of working-class society advantages similar to those which this Institute offers males' (*Huddersfield Institute Minutes of the General Committee* 1846: 8). The report went on to say that 'Huddersfield claims the pre-eminent distinction of having been the first to establish an institute organised and managed on a separate and independent basis for education of the young women of the working classes' (ibid.).

With the support from the male Institute in the town, where Schwann was President, the Female Institute offered 'sewing, reading, geography, history and other branches of a sound, moral, secular education … but no studies hostile to religion'. The teaching of religion was constrained to the church schools, and historically institutes had consciously made the decision not to debate religion or indeed politics (*Huddersfield Female Education Institute Minutes of the General Committee* 1847: 13). In 1848, 240 women attended classes in reading, writing, arithmetic, mental arithmetic, diction, grammar, needlework, singing and geography (*Huddersfield Female Education Institute Minutes of the General Committee* 1848).

At the first Annual Soiree held in 1848, Samuel Smiles, the guest of honour, spoke of his observations that in 'the district of Huddersfield I noted the number of women who signed the marriage registry with a cross was double that of men' (*Huddersfield Female Education Institute Minutes of the General Committee* 1848: 8) There had been some local opposition, including from the Secretary of the Huddersfield Mechanics' Institute, George Searle Phillips, to forming a female institute. The opponents were referred to as 'schemers' who tried to establish a strong opposition through scheming, which had little impact on the consensus of the populous of Huddersfield (*ibid.*: 25). By 1858, there were 46 classes and 597 students on the roll, clearly indicating that the Institute was very successful (*Huddersfield Female Education Institute Minutes of the General Committee* 1858: 2).

In 1883, however, it was agreed between the committees of the Mechanics' Institute and Female Education Institute to merge as to do so would provide the opportunity to expand membership, and under one roof. It seems to have been a relatively straightforward operation with little or no opposition to female students attending the larger male Mechanics' Institute, indicated by the March 1883 *Committee Minutes* stating that:

> the probable closing of the Female Institute and the availability of admitting females to all the privileges of the Mechanics' Institute were discussed at length. It was proposed and seconded and unanimously agreed that all facilities would be available to females as soon as the merger had taken place (*Huddersfield Mechanics' Institute Minutes* 6 March 1883: 21).

Less than two months later, a letter from the Secretary of the Female Institute confirmed that this had been a relatively easy transfer, with the initiative having come from the female Institute

> that the best thanks of the Committee be given to the Committee of the Huddersfield Institute for the readiness with which they accepted the suggestion to take over the work of the Female Institute hitherto carried on by this Institute for upwards of thirty-six years (*Huddersfield Mechanics' Institute Minutes* 1 May 1883: 34).

The decision to do this was both realistic and practical as the female Institute had not been as large as the male Mechanics' Institute and therefore did not have the same resources or opportunities. Unlike the male Institute, the female one had rented accommodation from a local school. By bringing both together in 1883 they supported what was to become a joint successful Technical School and Mechanics' Institute. However, separate classes for the females continued to be delivered and this reassured both present and future female students and their parents (*Huddersfield Mechanics' Institutes Minutes* 6 March 1883).

While Bradford and Huddersfield were the only towns with separate female institutes, and Leeds was unique in having large separate female-only classes in the male Mechanics' Institute in the Yorkshire Union, several others did take female students from 1850 onwards, albeit on a much smaller scale than men (see Table 6.3).

In the North East, there were several institutes supporting female membership. In 1853, the Middlesbrough Committee encouraged women to attend the lectures and classes on the same fee terms as male members (*Fifteenth Annual Report of the Yorkshire Union of Mechanics Institutes* 1853). Like state schools of

Table 6.3 Male to female membership at selective mechanics' institutes in the Yorkshire Union where data is available for 1850, 1861 and 1880

Mechanics' Institute	1850		1861		1880	
	M	F	M	F	M	F
Barnsley	219	37	267	40	506	70
Bingley	180	63	334	30	195	15
Boston Spa	120	20	No data	No data	83	63
Brighouse	130	11	133	4	206	3
Darlington	432	2	449	16	486	64
Dewsbury	105	0	358	34	415	61
Gomersal	61	39	191	39	79	7
Huddersfield	779	0	1226	17	1558	93
Huddersfield Female Institute	0	127	0	253	No data	No data
Lockwood	51	0	169	50	206	90
Keighley	319	101	400	30	1693	280
Leeds	1625	248	1201	268	No data	No data
Mossley	100	0	161	15	334	36
Pudsey	102	47	133	57	469	73
Selby	104	3	151	31	140	20
Sheffield	390	0	224	30	358	72
Thornton	100	18	98	50	194	6
Wakefield	522	25	835	35	688	75
Wentworth	74	2	78	6	50	10
Wilsden	144	56	78	12	57	13
York	494	12	439	9	620	90
Total	6051	811	6925	1026	8337	1141

Sources: *Twelfth Annual Report of the Yorkshire Union of Mechanics' Institutes* 1850; *Twenty-Second Annual Report of the Yorkshire Union of Mechanics' Institutes* 1860; *Forty-Second Annual Report of the Yorkshire Union of Mechanics' Institutes* 1880.

the later nineteenth century, mechanics' institutes segregated males from females as it was thought teaching should be carried out in single-sex classes. In 1859, at Middlesbrough there were three large classrooms, one for males, one for females and a third for juveniles (*Twenty-Second Annual Report of the Yorkshire Union of Mechanics' Institutes* 1860).

In 1857, at another North East institute, the Hartlepool West Institute, membership included ten women, and 31 youths (*Nineteenth Annual Report of the Yorkshire Union of Mechanics' Institutes* 1857). The Institute had also introduced an evening class for females during the winter of 1861 but this had not been successful with recruitment (*Twenty-Third Annual Report of the Yorkshire Union of Mechanics' Institutes* 1861). It seems that poor recruitment was due to the failure to find an appropriate female teacher, which was also the case at Middlesbrough in 1862 (*Twenty-Fourth Annual Report of the Yorkshire Union of Mechanics' Institutes* 1862). In 1873, the Darlington Railway Institute admitted female members at two shillings a year, as the Committee wanted to encourage family membership at classes and 'for them all to make good use of the library' (*Thirty-Seventh Annual Report of the Yorkshire Union of Mechanics' Institutes* 1875: 120). Redcar and Coatham Institute introduced a similar arrangement in 1882 (*Forty-Fourth Annual Report of the Yorkshire Union of Mechanics' Institutes* 1882).

In the Yorkshire Dales and Pennines, female membership was also encouraged. In 1848, for example, the Keighley Institute established a Female Improvement Society 'principally for the benefit of young women connected with our factories … an interesting [social] class, whose education hitherto has been much neglected' (*Tenth Annual Report of the Yorkshire Union of Mechanics' Institutes* 1848: 63). The Society offered reading, writing, arithmetic and grammar, as well as sewing and dressmaking, and during the first week 60 'young women enrolled and became full members', with membership rising to 140 later in the year. The Institute made specific mention of female education stating that 'questions are put to the classes, with a view of fixing the attention, exercising the intellect, and extending the boundaries of their knowledge' (ibid.). As a result of the success of the female improvement classes, a male improvement class was also introduced, in 1849, providing the 'same privileges to the young men who might feel disposed to repair the defects of a bad education' (*Eleventh Annual Report of the Yorkshire Union of Mechanics' Institutes* 1849: 65). Thus, Keighley Institute had identified the importance of elementary education for women at the same time as it was also being made available for men. It was in the era when institutes generally were responding to the needs of adult elementary education.

By 1862, the Keighley Committee was still continuing to question the benefit of women spending their days working in the factories with no opportunity to learn. It stated that 'the factory is not a place for females to learn household duties and if they have not obtained this knowledge in their youth, how can it be expected they will be able to make a home comfortable and happy!' The report went further:

> We should be glad to see two or three hundred gathered together on an evening, and taught plain sewing, and even plain cooking in the most economical manner. These females make the best wives who can make five shillings go the furthest with comfort (*Twenty-Fourth Annual Report of the Yorkshire Union of Mechanics' Institutes* 1862: 104–5).

This supports the work of Barker and Chalus (1997) and O'Day (2000), who saw the domestic work of women as being the main subject area in which to teach. The Committee was in effect stating that the Institute could provide a general elementary education for females, which their families would benefit from, and which would support educational development of their children as well as supporting their own mental improvement. In 1862, the Wilsden Institute Committee reported that there had been an increase in membership by 66 to 196 and this was due to new female members. In 1847 there had only been three females, but this rose to 67 as a result of the demand from women themselves that the Committee should reduce the membership costs to half that for men, and reduce the minimum age from 14 to 13 so that more females could attend (*Ninth Annual Report of the Yorkshire Union of Mechanics' Insitutes* 1847: 86).

Like Keighley, Bingley Institute also formed a Female Improvement Society in 1850 as more women wanted to become members, many of whom were 'employed in our factories'. In only a few months female membership had reached 60 (*Twelfth Annual Report of the Yorkshire Union of Mechanics' Institutes* 1850: 24). By 1860 there had been a considerable number of new female members and it was thought the introduction of a sewing class had encouraged many women to join (*Twenty-Second Annual Report of the Yorkshire Union of Mechanics' Institutes* 1860). The Committee had therefore responded positively to accepting females through offering classes they thought relevant to their needs.

In 1871, female members at Saltaire were given similar rights of access as males to the library and lectures, and the Committee hoped 'to be able to form some classes for their especial benefit during the winter months' (*Thirty-Third Annual Report of the Yorkshire Union of Mechanics' Institutes* 1871: 36). Three years later, the Committee reported that the female art classes were particularly popular. Girls from the recently established school, funded under the Education Act of 1870, joined these specific classes indicating that some institutes were taking school-age students for specialist classes.

Many of the mechanics' institutes offered female classes in elementary education. The reluctance of females to attend did not come from the institutes themselves, but from the women who were wary of attending what would have been male-dominated classes, a problem that Middlesbrough had responded to by offering separate classes for women. Once this had been identified, institutes set up separate classes, which increased membership and therefore income.

Mechanics' Institutes were ideally suited for working-class females as, like their male colleagues, classes were available in the evening. With higher wages in the northern textile towns, young women were drawn away from their families in the agricultural districts and gained their independence at a much earlier age than previously. Thus, there were large numbers of independent working women who could afford both the time and fees to attend their local mechanics' institutes or the separate ones in the case of Bradford and Huddersfield.

Despite the continuing demand made upon women by the manufacturing industries, particularly textiles, domestic service continued to expand during the nineteenth century and a new clerical sector for females began to replace male clerks. The invention and subsequent introduction of the typewriter saw office work as a less middle-class occupation. The introduction of the Workshop Acts of 1867 and elementary education, through the passing of the 1870 Education Act, and the establishment of compulsory education during the 1880s did much to reduce child labour (Price 1999: 45).

The Halifax Working Man's College had developed from the local mechanics' institute but was differentiated through its Christian Foundation similar to others, notably the original founded in London by Frederick Maurice (Harrison 1954). The Halifax College established a Young Women's Institute in 1860 and had three qualified female teachers. Women generally received education in the evenings, the same as their male counterparts, as they were working all day in the mills and factories (*Twenty-Second Annual Report of the Yorkshire Union of Mechanics' Institutes* 1860).

In 1856, Holmfirth introduced a female class but had had problems appointing competent teachers until the 'President had secured the gratuitous services of eight ladies' (*Eighteenth Annual Report of the Yorkshire Union of Mechanics' Institutes* 1856: 78). A female class of 18 students had been formed at nearby Honley in 1846, 'extending to the female portion of the community the means of an elementary education, and being sensible of the general neglect of this interesting portion of society' (*Eighth Annual Report of the Yorkshire Union of Mechanics' Institutes* 1846: 37). In 1874 Lindley Institute, on the outskirts of Huddersfield, had a membership of 90 males and 79 females. The Committee remarked that 'the Institute recognises the equal claims of girls and boys to a good education'. It also appreciated the support it got from employers and employees. There were several other institutes that offered female classes, providing evidence that there was support for working-class females. Others mechanics' institutes that had active female memberships included Elland, Shelley, Brighouse and Mirfield.

In 1857, at Lockwood Mechanics' Institute near Huddersfield, there were 212 males, who were offered courses in 11 subjects, and 45 females who were offered classes in 'reading, writing, arithmetic (as were the men), knitting, sewing and marking' (Purvis 1991: 38). The Ripon Institute Committee stated in 1863 that each student on the adult female courses paid 'tuppence' a week, 'one penny for use of the room, fire, gas, books, copy books, slates, and Library; the other penny was saved for the girl'. The second penny was saved weekly over the year and when a student left, often to enter service, the money was used to provide clothing for her new position. If a 'girl' withdrew from the course before the end of the year, the money was kept by the Institute. The Committee, no doubt from experience, recommended that all mechanics' institutes offer female classes separate from males to encourage more members (*Twenty-Fifth Annual Report of the Yorkshire Union of Mechanics' Institutes* 1863: 125).

With regard to female membership statistics, they are rather *ad hoc*, and only the years 1850, 1861 and 1880 provide the most informative data for several institutes listed in Table 6.3. The statistics have highlighted that Huddersfield Mechanics' Institute actually had 17 female members and 1226 males in 1860. The Bradford male Institute, in the same year, had 47 female members attending. It may be that those who attended the male institutes were following more advanced courses that were only offered in the mechanics' institutes.

Before 1850, few institutes had female students and this may have been due to reasons that Horrell and Humphries have identified as, 'most early nineteenth-century skills were readily learned; formal education was rare and irrelevant to female jobs; age-earning profiles were flat' (Sharpe 1998: 191). Purvis (1991) has identified from the 1851 Census that working-class female occupations included agricultural labourers, factory hands and domestic servants. She suggests that some bonnet-makers, seamstresses, dressmakers and washer-women were, in the main, also from this class. These categories help support the occupations of those who attended the female classes and, in particular, the Female Education Institutions at Bradford and Huddersfield.

The decisions at Bradford and Huddersfield to offer women their own institutions seem to have encouraged more to attend. This supports the claim that males and females did not like to be taught alongside each other in the case of elementary education, as was the case in primary school. Male teachers often taught the females separately. On the other hand, female-only curriculum, such as swering and home and hearth, were taught by females. However, in the case of all institutes that did have female members, committees encouraged their attendance in relation to additional fee income as well as offering them an education, whether it was elementary or 'home and hearth' courses. Gleadle (2001: 41) has suggested that women, in relation to those that lived in factory communities, were enthusiastic about taking up education opportunities. She states: 'the alacrity with which working-class women took advantage of adult education facilities indicates a widespread desire to improve intellectual attainments and a degree of self-respect and confidence'. This indicates that the mechanics' institutes, or at least some of those that were members of the Yorkshire Union, provided the opportunity for women to attend classes.

Conclusion

Mechanics' institutes welcomed female members and did respond to the needs of separate classes, female teachers and subjects relevant to female members, all of which encouraged women to attend in what was seen as a masculine culture. There was the additional factor that female membership contributed to the general success of institutes and brought in additional fees. However, it was only in 1873 that women were allowed to attend the Yorkshire Union of Mechanics' Institutes Annual Conference, which that year was held at Saltaire, near Bradford (*Thirty-Fifth Annual Report of the Yorkshire Union of Mechanics' Institutes* 1873). Berg (Sharpe 1998: 158) identified that 'apprenticeships, training or experience divided the wages of young females from the piece rates of older tradeswomen' during the industrial revolution. Davidoff and Hall (ibid.: 272) point out that the higher-skilled female trades such as millinery 'required apprenticeships with premiums as high as £50' to support these businesses. Such developments indicate that, as these financial rewards developed from having education and training, it became more broadly known that women would be more interested in an adult education.

Domestic education was only established formally through the Local Authorities in 1880. Such courses involved cookery, laundry-work and general housewifery, and attracted working-class females to attend. The Education Code of 1882 gave Local Authorities financial support to offer these courses, something mechanics' institutes had been doing since the 1850s. What increased the demand for such training was the improvement in housing, which 'raised standards in cleaning', and 'dietary diversification [which] was a notable feature of working-class life from the 1850s' (ibid.: 343).

Thus, these opportunities associated with female occupations were seen as skilled trades, so the demand from women to attend classes increased, the foundation of which was established by mechanics' institutes. This is reflected in the number of women, even if not substantial, who attended the Yorkshire Union institutes after 1850. The Female Institutions in Bradford and Huddersfield were able to concentrate on women-specific subjects. Students at both were mainly milliners, dressmakers and domestic servants during the 1840s, when both Institutes were founded. By the 1860s, the occupations were spinners, weavers and other factory hands (Purvis 1991: 38). The mechanics' movement raised the profile of female working-class education. It is beyond reasonable doubt that the mechanics' institutes, particularly in the Yorkshire Union, had a working-class female membership, albeit on a smaller scale to working-class males. The small numbers were not due to opposition; despite the 1851 census recording that 71 per cent of teachers in England and Wales were female, the main difficulties mechanics' institute committees had in offering classes for females was due to the lack of women teachers. It was not uncommon that classes for women often had to close through the lack of female teachers (Shoemaker 1998: 185–6).

References

Adams, F. J. (ed.) (1970) *Education in Bradford since 1870* (Bradford: Bradford Corporation).

Barker, H., and Chalus, E. (1997) *Gender in Eighteenth-Century England* (London: Longman).

Barker, J. (2010) *The Brontes* (London: Abacus).

Brown, V., and Inkster, I. (2000) 'Estimating a public sphere: Intellectual and technical associations at the time of the Great Exhibition', Inkster, I., Griffin, C., Hill, J., and Rowbotham, J. (eds) *The Golden Age: Essays in British Social and Economic History, 1850–1870* (Aldershot: Ashgate): 164–174.

Davidoff, L., and Hall, C. (1998) 'The hidden investment: Women and the enterprise', Sharpe, P. (ed.) *Women's Work, The English Experience 1650–1914* (London: Arnold).

Gleadle, K. (2001) *British Women in the Nineteenth Century* (Basingstoke: Palgrave Macmillan).

Harris, J. (1993) *Private Lives, Public Spirit: Britain 1870–1914* (London: Penguin).

Harrison, J. F. C. (1954) *A History of the Working Men's College 1854–1954* (London: Routledge and Kegan Paul).

Hertz, F. (1860) 'Mechanics' Institutes for working women, with special reference to the manufacturing districts of Yorkshire', *Transactions of the National Association for the Promotion of Social Sciences*, 347–354.

Horrell, S., and Humphries, J. (1998), 'Women's labour force participation and the transition to the male-breadwinner family, 1790–1865', Sharpe, P., and Arnold, E. (eds) *Women's Work: The English Experience, 1650–1914* (Cambridge: Cambridge University Press): 261–287.

Huddersfield Chronicle (12 June 1883).

Huddersfield Female Education Institute (1847) *Minutes of the General Committee.*

Huddersfield Female Education Institute (1848) *Minutes of the General Committee.*

Huddersfield Female Education Institute (1858) *Minutes of the General Committee.*

Huddersfield Female Education Institute (1862) *Minutes of the General Committee.*

Huddersfield Mechanics' Institute (1846) *Minutes of the General Committee.*

Huddersfield Mechanics' Institute (6 March 1883) *Minutes of the General Committee.*

Huddersfield Mechanics' Institute (1 May 1883) *Minutes of the General Committee.*

Hudson, P. (1992) *The Industrial Revolution* (London: Arnold).

Martineau, H. (1851) 'The history of England during the thirty years' peace', Watts, R. (1998) *Gender, Power and the Unitarians in England, 1760–1860* (London: Harlow).

O'Day, R. (2000) 'Women and education in nineteenth-century England', Bellamy, J., Laurence, A., and Perry, G. (eds) *Women, Scholarship and Criticism, Gender and Knowledge c. 1790–1900* (Manchester: Manchester University).

Price, R. (1999) *British Society, 1680–1880: Dynamism, Containment and Change* (Cambridge: Cambridge University Press).

Purvis, J. (1989) *Hard Lessons: The Lives and Education of Working-Class Women in Nineteenth-Century England* (Oxford: Blackwell).

Purvis, J. (1991) *The History of Women's Education* (Milton Keynes: Open University Press).

Rose, S. O. (1992) *Limited Livelihoods, Gender and Class in Nineteenth-Century England* (California: University of California).

Royle, E. (1987) *Modern Britain, A Social History 1750–1985* (London: Arnold).

Shacklady, E. (date unknown) *The Mechanics' Institute – Its Growth and Influence on the Cultural Life of Burnley,* unpublished local history article.

Sharpe, P. (ed.) (1998) *Women's Work, The English Experience 1650–1914* (London: Arnold).

Shoemaker, R. B. (1998) *Gender in English Society 1650–1850, the Emergence of Separate Spheres?* (London: Longman).

Skedd, S. (1997) 'Women teachers and the expansion of girls' schooling in England, c.1760–1820', Barker, H., and Chalus, E. (eds) *Gender in Eighteenth-Century England* (London: Longman): 105–125.

Tylecote, M. (1957) *The Mechanics' Institutes of Lancashire and Yorkshire Before 1851* (Manchester: Manchester University Press).

Watts, R. E. (1980) 'The Unitarian contribution to the development of female education, 1790–1850', *History of Education* Vol. 9, No. 4, 273–286.

Watts, R. (1998) *Gender, Power and the Unitarians in England, 1760–1860* (London: Harlow).

Yorkshire Union of Mechanics' Institutes (1846) *Eighth Annual Report.*

Yorkshire Union of Mechanics' Institutes (1847) *Ninth Annual Report.*

Yorkshire Union of Mechanics' Institutes (1848) *Tenth Annual Report.*

Yorkshire Union of Mechanics' Institutes (1849) *Eleventh Annual Report.*

Yorkshire Union of Mechanics' Institutes (1850) *Twelfth Annual Report.*

Yorkshire Union of Mechanics' Institutes (1852) *Fourteenth Annual Report.*

Yorkshire Union of Mechanics' Institutes (1853) *Fifteenth Annual Report.*
Yorkshire Union of Mechanics' Institutes (1856) *Eighteenth Annual Report.*
Yorkshire Union of Mechanics' Institutes (1857) *Nineteenth Annual Report.*
Yorkshire Union of Mechanics' Institutes (1858) *Twentieth Annual Report.*
Yorkshire Union of Mechanics' Institutes (1860) *Twenty-Second Annual Report.*
Yorkshire Union of Mechanics' Institutes (1861) *Twenty-Third Annual Report.*
Yorkshire Union of Mechanics' Institutes (1862) *Twenty-Fourth Annual Report.*
Yorkshire Union of Mechanics' Institutes (1863) *Twenty-Fifth Annual Report.*
Yorkshire Union of Mechanics' Institutes (1865) *Twenty-Seventh Annual Report.*
Yorkshire Union of Mechanics' Institutes (1871) *Thirty-Third Annual Report.*
Yorkshire Union of Mechanics' Institutes (1873) *Thirty-Fifth Annual Report.*
Yorkshire Union of Mechanics' Institutes (1875) *Thirty-Seventh Annual Report.*
Yorkshire Union of Mechanics' Institutes (1880) *Forty-Second Annual Report.*
Yorkshire Union of Mechanics' Institutes (1882) *Forty-Fourth Annual Report.*

7 From rented accommodation to civic pride

Introduction

It has already been implied that mechanics' institutes and similar societies were often established in rented accommodation. Rose (2001) points out that the mutual improvement societies, which had similar origins to those of mechanics' institutes, were relatively easy to set up by working men who supported a need for elementary education which the working classes wanted. Many of these became mechanics' institutes, among them Hebden Bridge, near Halifax, where the fustian cutter Joseph Greenwood was one of several men who founded a mutual improvement society in 1854. Greenwood, born in 1833, was the son of a handloom weaver and had had little formal schooling. However, he was greatly inspired as a result of reading Daniel Defoe's *Robinson Crusoe*; 'it opened up a world of adventure, new countries and peoples, full of brightness and change, an unlimited expanse' (Rose 2001: 108). Greenwood later wrote how 'renting an empty cottage with absolutely no furniture, we met and stood in a circle, one holding the candle while we deliberated and another wrote out the resolutions on paper' (ibid.). A year later there were 131 members and the Hebden Institute had outgrown the original premises. Most institutes started in a similar way.

One important indicator of the success of institutes was the development of the physical environment in which teaching and learning took place. Committees measured their success not only by the number of members and the volumes in their libraries, but also in their buildings. There were times when the success of an institute, as at Hebden Bridge, meant it outgrew its original rented property.

Institutes identified the need to canvass for new members to increase membership and fee income for the purpose of establishing new rented buildings and for upgrading established ones. At Darlington, for example, the Committee was concerned that 'the advantages which the Institution affords are not sufficiently appreciated' and they hoped members would canvass throughout the area (*Twelfth Annual Report of the Yorkshire Union of Mechanics' Institutes* 1850: 32). Hartlepool West ran parents' nights in 1857 to encourage them to send their children and themselves to become members, and the result was that 40 new members between the ages of ten and 30 enrolled at the Institute (*Nineteenth Annual Report of the Yorkshire Union of Mechanics' Institutes* 1857: 76). Mechanics' institutes, when funds allowed, were always keen to extend and improve their accommodation, and some were rebuilt and resited more than once. In the North East, the Lingdale Miners' Institute Committee in 1885 believed that if the Institute had better premises then it would attract a larger membership and therefore more income (*Forty-Fourth Annual Report of the Yorkshire Union of Mechanics' Institutes* 1882: 110).

In the Dales and Pennines alone, 18 Yorkshire Union institutes had accommodation issues that were serious enough to discuss in their annual reports. In some cases buildings were funded directly by their patrons, such as the industrialists Edward Pease, who provided new accommodation for Darlington, and Sir Titus Salt who did the same at his model village, Saltaire. In most cases, however, committees relied on private donations, for example from W. E. Forster who donated £5 to the building fund of both the Otley and Burley institutes near Bradford, as well as income from membership, public lectures, events and exhibitions. The popularity of the Institute at Keighley, for example, caused severe pressure on the existing accommodation. The membership established a Building Committee in 1865 to put together plans and identify a location for a new Institute, supported by the Duke of Devonshire's estate.

In 1866, the Keighley Committee established a building fund for an Institute. Important considerations in planning the new building were that there must be a lecture hall, library, reading rooms and accommodation for evening classes. There was also to be space for a Science and Art School, as approved by the Science and Art Department in Whitehall, as well as a Trade School, which 'should supply to all classes, from artisans upwards, a liberal and practical English education, supplemented by the systematic teaching of such art and science subjects as are applicable to the trades of the district' (*Thirty-Third Annual Report of the Yorkshire Union of Mechanics' Institutes* 1871: 27). The term 'school' was used to mean department.

In 1871, the new building at Keighley was opened at a cost of £15,000 (see Figure 7.1).

There was very little money in the building fund due 'to the monetary crisis and dullness of trade' and it would be several years before the debt was paid off (ibid.: 29). Despite the continuing building programme at Keighley, the Institute was still expanding faster than accommodation would allow, even with the extra space of the new wing, which was opened in 1889 and partly funded by the Worshipful Company of Clothworkers (*Annual Report of the Yorkshire Union of Mechanics' Institutes* 1889). By 1890,

the extensive accommodation supplied by the new wing is now all utilised to the fullest capacity of the building, and if there were more class-rooms they could be filled, so great is the response to the efforts of the Council [Committee] and teachers to attract students to the Institute (*Fifty-First Annual Report of the Yorkshire Union of Mechanics' Institutes* 1890: 114).

Another institute with accommodation difficulties was Skipton, which by 1852 had a shortage of space. In 1863, 11 years later, the recently constructed Town Hall in the town was being used as an overspill for the Institute to hold a series of

Figure 7.1 Keighley Mechanics' Institute and Technical School.

Source: *Thirty-Ninth Annual Report of the Yorkshire Union* 1877.

public lectures, readings and musical evenings. There were often 600 attendees at these weekly events, generating additional income and contributing to a building fund (*Fourteenth Annual Report of the Yorkshire Union of Mechanics' Institutes* 1852).

As late as 1882, the accommodation problem had still not been solved, with the continual increase in the number of members at Skipton. Some evenings' classes had doubled up and there were particular difficulties with the elementary classes, which were 'overcrowded with boys' (*Forty-Fourth Annual Report of the Yorkshire Union of Mechanics' Institutes* 1882: 104). If more space was available, then girls could also receive an elementary education. The Committee noted that 'there is urgent need for extended class-room accommodation for the youths of Skipton, but the necessity for giving equal educational facilities to the girls is quite as pressing'. The Institute did accept females and there were 'considerable numbers ... entered the Science and Art Classes this year' (ibid.: 149).

The incentive for raising funds to build a new institute at Skipton to cope with expansion of members came from Bradford, which had built a large institute at a cost of £32,000. The Committee at Skipton were encouraged to raise funds for a new building as a result. A grant of £829 came from the Science and Art Department, and the new building was finally opened in 1894. The frontage included two shops which provided rents and income to support the institute (Carter and Weatherhead 1999). The total cost was £5513 (*Skipton New Mechanics' Institute and Science and Art Schools Building Committee Statement* 1894: 1). Skipton Institute went on to be one of many successes, and in the twentieth century it became a further education college with the original 1894 building still used as one of its centres.

The same was true in other areas. The rented buildings of the Huddersfield Institute, for example, were not large enough for the number of members it was attracting. In 1857, the Society of Arts examinations for the North of England took place in the Riding School where 248 students from several institutes, including Huddersfield, sat examinations, as the Institute's building was not large enough (*Huddersfield and Holmfirth Examiner* 1 August 1857). By 1859, a building fund for a new Institute at Huddersfield had raised £3700 from bazaars and annual soirees. Workmen in one mill alone had raised £30 during the year in support of larger accommodation. A further £1200 was required before building work could commence, replacing four separate rented buildings that were costing, between them, £80 a year (*Twenty-First Annual Report of the Yorkshire Union of Mechanics' Institutes* 1859: 92).

A new purpose-built Institute was built in 1861 in Northumberland Street (*Twenty-Fourth Annual Report of the Yorkshire Union of Mechanics' Institutes* 1862). However, within eight years it had outgrown the building. In 1869, the Committee reported to the Yorkshire Union that the textile classroom was 'used for the application of design to the loom, and every seat is filled'. For the first time since practical chemistry had been introduced, the maximum number of 15 students had been reached and others had to be put on the waiting list. Further accommodation was therefore required so that the institute could continue to develop and expand (see Figure 7.2).

The plans for a second, larger Mechanics' Institution and Technical School at Huddersfield were agreed at a Joint Committee of the male and female Institutes in 1881. They had identified the need for a new building which could accommodate males and females. as well as supporting further developments of various classes including those connected with textiles and other trades of the district. Far better accommodation was also required for the School of Art and the Science School, as well as for a penny bank, reading room and larger library (*Forty-Third Annual Report of the Yorkshire Union of Mechanics' Institutes* 1881: 33). The new building was financially supported by the Worshipful Company of Clothworkers, and the Schwann and Ramsden families, along with private donations. The building was opened in 1884 (*Huddersfield Chronicle* 18 June 1884). In 1886, it became the Technical College.

The Huddersfield Female Education Institute also had a shortage of accommodation caused by demand, which was highlighted in the Committee's report of

Figure 7.2 Huddersfield's first purpose-built Mechanics' Institute, located in
Northumberland Street, was built in 1861 and was basic in design.
Within a few years a building fund was set up to replace it with a much
larger building.

Source: Author's collection.

1860: 'In consequence of not having a room sufficiently large in which to collect
all the pupils of the Institution, the committee have been unable to make those
arrangements for lectures and addresses they could have wished' (*Twenty-Second
Annual Report of the Yorkshire Union of Mechanics' Institutes* 1860: 6) The accommodation at the time was located in the Netherwood's Buildings in King Street
(*Twentieth Annual Report of the Huddersfield Female Education Institute* 1858:
2). By 1869 the Institute had moved into the Gladstone Buildings, which were
more suitable, and by 1875 it had moved for the final time into the Board School
on Beaumont Street, never actually owning its own building (*Thirty-Seventh
Annual Report of the Huddersfield Female Education Institute* 1875). In 1883 the
Female Institute moved into the new Technical School and Mechanics' Institute,
which it shared with its male counterpart, and the two institutions became one.

The Brighouse Mechanics' Institute, near Huddersfield, also had accommodation problems by 1848, having to rely on a religious establishment that was
made available during the week (*Eleventh Annual Report of the Yorkshire Union
of Mechanics' Institutes* 1849: 29). In 1853, the Institute moved into new premises; 'the removal of the Institute to new and more commodious rooms during
the year, has been attended with good effects, as evinced by the large increase of
members of that class for whose benefit such institutions are principally intended'
(*Fifteenth Annual Report of the Yorkshire Union of Mechanics' Insitutes* 1853). In
the meantime, under a memorandum of association, Brighouse Town Hall provided rooms for some classes, a library, reading rooms and newsrooms, for a fee.
There is no evidence that the Institute moved again, having to make do with this

building and additional classes in the town hall (*Twenty-Ninth Annual Report of the Yorkshire Union of Mechanics' Institutes* 1867: 23).

The Greetland and West Vale Institute, near Halifax, sent in its first report to the Yorkshire Union in 1873, although it had been founded in 1847, suggesting that previously it had not been a member of the Union. A new, purpose-built Mechanics' Institute was about to be opened in 'this thickly populated manufacturing district', responding to the growth in number of students. There was also a shop and some dwellings included in the building (see Figure 7.3), which, as already highlighted, would bring in much-needed income through rents (*Thirty-Fifth Annual Report of the Yorkshire Union of Mechanics' Institutes* 1873: 81).

Shortage of accommodation was also a serious problem at Bradford, Bridlington Quay, Halifax, Leeds, Ripon and York. In the case of the latter, for example, York Institute started in a small house in Bedern in the centre of the old city in 1827 but a year later it had moved into larger premises in St Saviourgate, the premises of which were extended in 1846 at a cost of £765. In 1883, new large premises were opened in Clifford Street, at a cost of £7,500, to a twelfth-century French design (Gardiner 1991: 4).

There are one or two examples where purpose-built institutes were established in rural communities with small numbers. One such example was at Brough, near Hull, in Yorkshire. Brough was the smallest and most basic purpose-built institute in the Yorkshire Union of Mechanics' Institutes, if not the whole country. It was built in 1878 to a design similar to that of prefabricated chapels such as

Figure 7.3 Greetland and West Vale Mechanics' Institute in the West Riding of Yorkshire was built in 1874.

Source: *Thirty-Ninth Annual Report of the Yorkshire Union of Mechanics' Institutes* 1877.

those often built for navvy railway communities or in towns with an expanding population (see Figure 7.4).

However, for other institutes there was not the need to move out of rented accommodation, due to small membership numbers, such as at the one at Oakworth. For others, purpose-built accommodation developments were ill-timed, such as at Kettlewell, where no sooner had new accommodation been provided than the population began to leave, looking for more regular work in the growing towns, which resulted in a fall in membership.

With the introduction of government examination grants and sponsors, such as those proved by the Worshipful Company of Clothworkers (see Table 7.1) and the Science and Art Department, as well as increases in membership, larger institutes were able to build or expand their accommodation, in some cases more than once.

There were other mechanics' institutes and technical schools not listed specifically in the *Trusts and General Superintendence Minutes* for 1875 to 1914 but that were under the general heading of the Yorkshire Union of Mechanics' Institutes. Among them was Embsay Institute, near Skipton, which was supported to the amount of £5 for a prize for the best design on fabric in 1879 (*Thirty-Seventh Annual Report of the Yorkshire Union of Mechanics' Institutes* 1875: 98).

Civic pride

One of the important underlying developments identified has been the importance the committees of institutes put on having both enough classroom space and appropriate accommodation. When financial conditions allowed, there was eagerness amongst committees to build institutes and to be seen as part of the general public building programme of the mid- and late nineteenth century.

Figure 7.4 Brough Mechanics' Institute near Hull in Yorkshire.

Source: *Forty-Second Annual Report of the Yorkshire Union of Mechanics' Institutes* 1880: 23.

Table 7.1 List of the former Mechanics' Institutes supported and recognised by the Worshipful Company of Clothworkers, London

Institution	Date first mentioned in records
Yorkshire College of Science (later University of Leeds)	1875
Bristol College of Science	1875
University College Bristol	1877
*Huddersfield Mechanics Institute and Weaving School, Yorkshire	1877 and 1878
Glasgow Weaving School and Glasgow Technical College, Scotland	1877
**Yorkshire Union of Mechanics' Institutes	1877
Bradford Technical Weaving School, Yorkshire	1878
Barnsley, Yorkshire	1878
Hull, Yorkshire	1878
Keighley Mechanics Institute and Trade School, Yorkshire	1878
Batley Technical School, Yorkshire	1878
City and Guilds of London Institute	1879
Saltaire Mechanics Institute and School of Art, Yorkshire	1880
Salt Schools, Saltaire, Shipley, Yorkshire	1880
Kings College London School of Fine Art	1881
Keighley Weaving School, Yorkshire	1882
Harris Institute, Preston, Lancashire	1882
Bolton Technical Classes, Lancashire	1883
Dewsbury Technical School, Yorkshire	1883
Halifax Technical School, Yorkshire	1883
Macclesfield, Cheshire	1883
Bingley Technical School, Yorkshire	1884
Bristol Trade Schools	1885
Hawick, Scottish Borders	1886
Morley Technical School near Leeds, Yorkshire	1886
Leicester, Leicestershire	1886
Onslow College of Science, Hatfield, Hertfordshire	1886
Blackburn Technical School, Lancashire	1887
Victoria University, Manchester, Lancashire	1888
Durham College of Science, County Durham	1888
Dundee Technical Institute, Scotland	1888
Galashiels Science Schools, Scottish Borders	1888
Nicholson Institute, Leek, Staffordshire	1889
Ossett Technical School, Leeds, Yorkshire	1889
Islington Polytechnic, London	1889
Wakefield Technical School, Yorkshire	1890
Stroud Technical School (had been supported before as part of University College Bristol), Gloucestershire	1891
Trowbridge Technical School, Wiltshire	1891

(continued)

Table 7.1 Continued

Institution	Date first mentioned in records
Westbury and Whorwellsdown, Wiltshire	1891
Holmfirth Technical Institute, Yorkshire	1893
Swansea, South Wales	1893
Guisburn Institute, Lancashire	1893
Cleckheaton School of Science and Art, Yorkshire	1894
Bradford on Avon Textile and Technical School, Wiltshire	1895
Cowper Street Commercial Classes, London	1895
Hackney Technical Institute, London	1896
Lindley Mechanics' Institute Yorkshire	1896
Congleton Technical Institute Cheshire	1897
University College of Wales Aberystwyth, Wales	1899

Source: *The Worshipful Company of Clothworkers' Trusts and General Superintendence Minutes* 1875–1914.

Notes

* A donation of £1,000 for the new Huddersfield Institution was given by the Worshipful Company of Clothworkers, who visited in June 1880 and had been impressed with the quality of teaching. (*Huddersfield Chronicle* 1880).

** The Worshipful Company of Clothworkers donated 100 copies of *The Dyeing of Textile Fabrics* by J. J. Hummell in 1885 to the Yorkshire Union, for the circulating library (*Forty-Seventh Annual Report of the Yorkshire Union of Mechanics' Institutes* 1885: 94).

Harrison (1961: 213) identified that from the 1860s 'new mechanics' institute buildings were opened with due civic ceremony'. The unnamed author in the *Builder* (1850), making reference to institutes, wrote 'it is a very evident fact that a beautiful building goes a long way in the adornment of a society with the character of responsibility and importance'. St John (1858: 221), the Radical journalist and traveller, stated that 'in calculating the effect of mechanics' institutions on the students who frequent them, we must take into account the architectural character of the buildings themselves, which often possess an air of imposing grandeur'. By the second half of the nineteenth century elaborate buildings were being erected, such as the one at Newcastle-upon-Tyne, which was described by the *Builder* (1865: 304) thus:

> the new building, which has been designed by Mr Thomas Oliver, architect, stands on a piece of ground in New Bridge Street ... the style is Italian. The interior will consist of six large sized classrooms, a library, a lecture room, a news room, a 'smoke room' and an extensive corridor leading from the entrance to the grand staircase.

Later institute developments were often located near town halls and other public buildings, and were designed in the Neo-Gothic style of municipality influenced by the High Victorian movement in architecture covering the period 1850 to

1870 (Muthesius 1972). It was Augustus Pugin (1812–1852), the English architect, designer, artist and critic, who is chiefly remembered for his pioneering role in the Gothic Revival style, culminating in the interior design of the Palace of Westminster. Sir George Gilbert Scott (1811–1878), was also an English Gothic Revival architect, chiefly associated with the design, building and renovation of churches and cathedrals, although he started his career as a leading designer of workhouses. (Cole 1980: 1). Edward Hughes, a protege of Scott, designed Huddersfield Mechanics' Institute and Technical School, which was opened in 1883. Hughes was also the designer of the town's Market Hall, bank and Spring Grove School. The Technical School cost £20,000 to build and its ornate facade is adorned by lions holding shields that bear the coats of arms of Sir Thomas Brooke, President of the Mechanics' Institute at the time and Chairman of the Chamber of Commerce, Huddersfield Corporation, Sir John William Ramsden, landowner, and the Clothworkers' Company; the latter donating £1,000 to the building fund (Walker and Haigh (2009). Pugin's style quite evidently influenced schools and institutes' Gothic Revival styles, although the only educational establishments he seems to have designed were the University of Glasgow, the Vaughan Library at Harrow Public School in London and the Albert Institute in Dundee, Scotland. However, his influence on local architects with regard to public building can be seen through hospitals, town halls, schools, libraries and mechanics' institutes (Muthesius 1972).

Stobart (2004: 485) sums this up well. 'Civic culture, the construction of town halls, museums, libraries, concert halls, and the like – was central to identity and image of both the town and its elite'. By the mid-nineteenth century a 'new network of voluntary associations and public institutions aimed at cultural and moral improvement' was being planned and built in many towns, including the 'foundation of learned societies and the construction of libraries and newsrooms'. There was an 'emphasis on public visibility and display, on ritual and performance, and on select conviviality'. Above all, 'this was reflected in the increasingly grandiose public buildings'. In many towns, post–1850, purpose built mechanics' institutes contributed to this 'cultural and moral improvement'. Stobart also suggests that Victorian civic culture was an instrument of social control. It was, he believes, a concerted effort by the middle-class 'to (re)create the towns and cities of industrial Britain in their own image'. They believed that it 'would help to create model citizens – rational, self- improving and public spirited – imbuing the working classes with middle-class values and bringing them within the city as a social construct and social ideal'. It was in 'this spirit that we see the spread of mechanics' institutes, and of public parks, town halls, libraries and theatres' (ibid.: 487). Accrington, Bradford, Huddersfield, Leeds, Manchester, Wakefield and York are all examples of institutes that were located near town halls and other civic buildings.

These later purpose-built mechanics' institutes were often influenced by a variety of architectural styles based on chapels, schools and, in the case of Dewsbury, Huddersfield, Keighley, Pudsey, Yeadon and York, their architects were influenced by the Gothic School of the second half of the nineteenth century. This 'quintessential style of the Middle Ages was revived through church building, quasi-monastic colleges of Oxford and Cambridge and government and civic buildings, such as town halls and mechanics' institutes' (Bradley 2007: 23). Others were influenced by school board designs, such as those at New Marske and Slaithwaite, and Nonconformist chapel designs, such as the one at Eccleshill near Bradford.

Contribution to the civic landscape was another important indicator of the success of institutes. Committees measured their success not only by the number of members, and the volumes in their libraries, but also in the development of the physical environment.

Bradford Mechanics' Institute's first building was established in 1839, relying on rented premises (*First Annual Report of the Yorkshire Union of Mechanics' Institutes* 1839: 128). A building fund was started to raise the £9500 required for a new larger Institute, and the industrialists Salt and Ripley contributed to it (*Twenty-Ninth Annual Report of the Yorkshire Union of Mechanics' Instiutes* 1867: 129). In 1871, a new mechanics' institute was built in what was then referred to as the 'Bradford-Italianate style', at a cost of £23,500. The town's

> elite built it "in the Venetian style, adding symbols of their own history to those of the historic trading empire". In this way, the construction and rhetoric of civic culture could simultaneously link the past with the present, modernity with morality, industry with the arts (Stobart 2004: 486).

The Mechanics' Institute, just one of several buildings constructed in this style at Bradford, was opened by W. E. Forster, sometime President of the Bradford Mechanics' Institute. The new building could accommodate 700 students and the lecture room held 1500. The library had 9935 books (Adams 1970: 203).

There is no doubt that in many cases the shortage of space resulted in restrictions of growth in membership, and not just in the larger towns. Many institutes relied on donations and income from galas and annual soirees to establish a building fund, as well as increased fees from growing membership. The Worshipful Company of Clothmakers supported several institutes in the woollen districts of Yorkshire, such as Bradford, Huddersfield, Keighley and Leeds. As in other sectors, the increase in the size of accommodation and often purpose-built institutes reflected the success of the Yorkshire Union and the movement as a whole.

Pudsey Mechanics' Institution had periods of fluctuating membership but became an established Institute during the second half of the nineteenth century, as reflected in the style of the 1880 building. It was located in the centre of the town and in 1912 it became the town hall (see Figure 7.5).

Yeadon's Mechanics' Institution and Town Hall was of a similar municipal Gothic design to many others, including the ones at Keighley and Pudsey (see Figure 7.6).

It was common practice to build later institutes, both in the Yorkshire Union and across the British Isles, close to the town hall, reinforcing the relation between local government and education. The new and final Mechanics' Institute and

Figure 7.5 Pudsey Mechanics' Institute.

Source: *Forty-Second Annual Report of the Yorkshire Union of Mechanics' Institute* 1880.

Figure 7.6 Yeadon's Mechanics' Institute.

Source: *Sixty-Fourth Annual Report of the Yorkshire Union of Mechanics' Institutes* 1902: 126.

Technical School was opened in 1884. Later, the building was used Huddersfield Technical College and is now part of the University of Huddersfield's School of Human and Health Sciences (see Figure 7.7).

The new purpose-built institutes allowed for up-to-date teaching accommodation and laboratories such as those at Huddersfield, which included both chemistry and dye laboratories (see Figures 7.8 and 7.9).

The Exhibition of 1883 to raise money for the Huddersfield Institute included a machine hall relating to the textile industry. Many of these were donated to the Institute, as were subsequent examples, so students could be trained operating them (see Figure 7.10).

Leeds Mechanics' Institution and Literary Society was the second-largest in Britain. It was designed by Cuthbert Broderick (1821–1905), who won a £200 competition to build Leeds Town Hall, located nearby. He also designed the Corn Exchange in the town. The building, which cost £20,000 and took five years to construct, was opened in 1865 (*Daily Mail* 9 November 2007: 58). It became part of the Leeds College of Art in 1903, which was extended at the back. Later the Institute became a theatre and is now the city's museum. Leeds Mechanics' Institution was also the Headquarters for the Yorkshire Union (see Figure 7.11).

New Marske Mining Institute, a member of the Yorkshire Union, was built and funded by the Pease Mining Company and the design was influenced by Victorian school architecture. The Institute comprised classrooms, a library, a

Figure 7.7 Huddersfield Mechanics' Institute and Technical School of 1884.

Source: Author's collection.

Figure 7.8 The chemistry laboratory at Huddersfield Technical School and Mechanics' Institute 1881, complete with blackboard and fume cupboard.

Source: Heritage Quay, The University of Huddersfield Archives.

Figure 7.9 The dyeing laboratory at Huddersfield Technical School and Mechanics' Institute 1881.

Source: Heritage Quay, The University of Huddersfield Archives.

reading room and committee room. The small building on the left was the care-taker's house (see Figure 7.12).

Like New Marske, Slaithwaite Mechanics' Institute was built on a school board design and supported the education of children, as well as adults, following the passing of the 1870 Education Act (see Figure 7.13).

Figure 7.10 A selection of weaving machines at the Huddersfield Technical School and Mechanics' Institute in 1883.

Source: Heritage Quay, The University of Huddersfield Archives.

Figure 7.11 Leeds Mechanics' Institution and Literary Society.

Source: *Forty-Ninth Annual Report of the Yorkshire Union of Mechanics' Institutes* 1887.

Figure 7.12 New Marske Mining Institute.

Source: *Thirty-Ninth Annual Report of the Yorkshire Union of Mechanics' Institutes* 1877.

Figure 7.13 Slaithwaite Mechanics' Institute.

Source: *Forty-First Annual Report of the Yorkshire Union of Mechanics' Institutes* 1879.

Prior to the 1870 Elementary Education Act, village schools built from the late 1840s were often of Gothic design. Architects of mechanics' institutes were often influenced by school designs similar to those at the New Marske Miner's Institute and Slaithwaite Mechanics' Institute. Publically funded board schools, following the passing of the 1870 Act, were often built in densely-populated areas and were therefore not always single-storey, unlike the above, which were located in spacious newly developed communities (Dixon and Muthesius 1978).

In 1828, York Mechanics' Institute was founded and rented a building in the former medieval centre in St Saviourgate. The late nineteenth-century York Mechanics' and Scientific Institute was built in Clifford Street where the town hall, police and fire stations were also built at about the same time (see Figure 7.14). It was opened in 1883 at a cost of £7500, to a twelfth-century French design (Gardiner 1991: 4).

Cleckheaton Mechanics' Institute, near Brighouse, was founded in 1838, by the local mill owner George Anderton and his son William. It became a member of the Yorkshire Union in 1840 (Hird 1985: 41). Cleckheaton had a very

Figure 7.14 The York Mechanics' and Scientific Institute.

Source: *Forty-Fourth Annual Report of the Yorkshire Union of Mechanics' Institutes* 1882.

strong Nonconformist communities particularly Unitarians, may have influenced chapel designs similar to the one at Calverley (see Figure 7.15).

When new larger purpose-built institutes were established, not only was teaching accommodation more in keeping with the needs for technical education but also the libraries were able to hold larger stocks of books and periodicals. They also provided the opportunity for 'penny banks' to be set up within the institutes.

Savings banks, sometimes referred to as penny banks, were seen as institutions which rewarded thrift and therefore reduced dependence in old age on the Poor Law. 'The industrious would be rewarded for their good habits and the idle and vicious would suffer if required to rely on their own exertions rather than on public assistance' (Moss and Russell 1994: 10–14). Penny banks were established to encourage the working classes to save a few pence, with interest, when they could, in order to support themselves financially. Samuel Smiles described such banks as:

> the poor man's purse. The great mass of the deposits are paid in sums not exceeding sixpence, and the average of the whole does not exceed a shilling. The depositors consist of the very humblest members of the working class, and by far the greatest number of them have never before been accustomed to lay by any portion of their earnings (Alter, Goldin *et al.* 1994: 735).

Such arrangements were very much associated with 'self-help' and with the Temperance Movement generally and on which mechanics' institutes were established. The Penny Bank Movement was first established in Glasgow and then spread across the British Isles and overseas (Ross 2002: 22).

According to Charles Dickens, Huddersfield was 'that which first called any large share of attention to the subject', that is, a penny bank which was established in 1850 and, relevant to this book, closely related to the town's Mechanics

Figure 7.15 Calverley Mechanics' Institute.

Source: *Thirty-Sixth Annual Report of the Yorkshire Union of Mechanics' Institutes* 1874.

Institute (ibid.: 27). Dickens believed that promoters of penny banks should 'graft it on some stable and successful institution for the working classes which may happen already to exist' (Dickens 1852: 424–5).

In 1850 Charles Sikes, who was manager of the Huddersfield Banking Company, wrote to Edward Baines, President of the Yorkshire Union of Mechanics' Institutes, putting forward a proposal that would support some of the issues highlighted by the Committee. He proposed a Savings Bank Committee which would open a room one night a week in Institute if they wished to do so, as working men were unable to attend during the day. Sikes insisted that where a working man was able to visit a savings bank, he was 'often regarded with disdain or even discourtesy if he presented the minimum deposit of one shilling'. The mechanics' institute savings bank, on the other hand, would encourage the working man to save, knowing that he would not be embarrassed if he put into his account a few pennies at a time and this would 'encourage habits of thrift, economy and forethought' (*Huddersfield Examiner* 29 April 1882).

The members would be able to inspect their balance twice a year and when they had two guineas saved it would be possible to transfer from the mechanics' institute savings bank to the independent one in the town. Not only did this reduce any embarrassment in saving small amounts of money, but there would be little risk to the institute itself. Such responses to the idea of integrating savings banks into mechanics' institutes showed a vision for diversity of business (*Twelfth Annual Report of the Yorkshire Union of Mechanics' Institutes* 1850).

By 1851, the savings bank at the Huddersfield Mechanics' Institute (which was, like many others, known as the 'penny bank' so as not to be confused with the town's saving bank) had 302 accounts, increasing to 350 in 1852. By 1875, the number of savers had risen to 2962. While this was the case at Huddersfield, it would soon after be common for mechanics' institutes and similar institutions to have their own penny banks. Working-class females were saving as well as men at penny banks and savings banks (Maltby 2011). The Yorkshire Union Committee met early in 1851 and considered:

> the practicality and desirableness, or otherwise, of establishing, on pure business principles and on the basis of the safest scientific computations, in connection with the Union and the several Societies forming it, a Mutual Insurance Fund, comprising the several departments of a Saving Bank, Sick Fund, Life Assurance, Deferred Annuity, Endowment of Children and other departments of prudential assurance, or any of them (*Twelfth Annual Report of the Yorkshire Union of Mechanics' Institutes* 1850: 41).

Several mechanics' institutes in the Yorkshire Union, particularly the larger ones, went on to run savings banks which were both popular and successful. While the Union reminded member institutes of their main responsibility of supporting education, they saw the introduction and development of the savings banks as contributing to the habit of members to save some money from their wages. The Union reminded the committees that savings banks must not become discredited as this would affect the individual institutes and the movement as a whole. The spread of the savings banks through many mechanics' institutes in Yorkshire resulted in the establishment of the West Riding of Yorkshire Providence Society and Penny Savings Bank on 17 November 1856. From this developed the Yorkshire Penny Bank, which was opened on 1 May 1859 (Popple 1991: 104). The Yorkshire Penny Bank had very strong ties with the Yorkshire Union of Mechanics' Institutes, as listed in Table 7.2.

Access to penny banks was made possible by institutes opening their doors to attracting their members to save as part of the Victorian belief in 'self-help'. The working classes, both male and female, who attended mechanics' institutes or similar, were not only supported with education and training, but also to save money rather than rely on the charity of the Poor Law. Without larger memberships, only possible in many cases with the expansion of accommodation, it seems likely that penny banks would have not been practical and members would have had to go to the savings banks in the growing towns.

Conclusion

In Chapter 1 the debate amongst many historians has been that the Mechanics' Institute Movement was a failure. Research has shown that many of them were

Table 7.2 Branches of the Yorkshire Penny Bank and other associated savings banks affiliated with the Yorkshire Union Institutes

Yorkshire Penny Banks

Branch	Branch	Branch
Adwalton	Halifax	Ossett
Appleton Wiske	Haworth	Otley
Barnoldswick	Helmsley	Ripon
Batley	Holme Lane	Saltburn-by-the-Sea
Bishop Monkton	Hunmanby	Scarborough
Castleford	Keighley	Sheffield
Dewsbury	Killinghall	Skipton
Eccleshill	Kirkby Malzeard	Slaithwaite
Garforth	Lascelles Hall	Wakefield
Gargrave	Leeds Working Men's Hall	
Gomersal	Lindley	

Other institute Penny Banks but not with the Yorkshire Bank

Bingley	Gargrave	Longwood
Bolton Abbey Group	Heckmondwike	Middlesbrough
Brighouse and Rastrick	Huddersfield	Milnsbridge
Cottingley	Keighley	Otley
Darlington	Kildale	Shelley
Darlington Albert Hill Institute	Lockwood	Wilsden

Source: *Annual Reports of the Yorkshire Union of Mechanics' Institutes.*

short-lived but often reopened, particularly when they responded to the needs of the working class. Not only do membership increases and the introduction of technical education and examinations indicate success during the second half of the nineteenth century, but also that many of the institutes moved into larger rented accommodation as they became successful. For some, they would always make use of rented buildings, but many raised funds to build their own institutes; some, notably Huddersfield, actually built a second, larger, Gothic-style building, reflecting the ongoing success. The purpose-built institutes became technical schools, schools of art and design, free public libraries or museums. Once the institute had its own premises, it could adapt the accommodation to include purpose-built classrooms and laboratories, a reading room separate from the library and a penny bank, where members could attend weekly to save money, with interest. Some institutes also had recreation facilities, such as gymnasiums. Their location, often in the town centre, near the town hall, indicates the importance of the institutes in providing education and social amenities in the second half of the nineteenth century.

References

Adams, F. J. (ed.) (1970) *Education in Bradford since 1870* (Bradford: Bradford Corporation).

Alter, G., Goldin, C., and Rotella, E. (1994) 'The savings of ordinary Americans: The Philadelphia Saving Fund Society in the mid-nineteenth century', *Journal of Economic History* Vol. 54, 735–767.

Bradley, S. (2007) *St Pancras Station: Wonders of the World* (London: Profile Books).

Builder (16 November 1850).

Builder (29 April 1865).

Carter, S., and Weatherhead, A. (1999) *Craven College, A History of Further Education in Skipton* (Skipton: Craven College).

Cole, D. (1980) *The Work of Sir Gilbert Scott* (London: The Architectural Press).

Daily Mail (9 November 2007). Answers to correspondents compiled by George Legge, p. 58, referring to the works of Cuthbert Broderick.

Dickens, C. (24 January 1852) *Household Words: A Weekly Journal* Vol. 4.

Dixon, R., and Muthesius, S. (1978) *Victorian Architecture* (London: Thames and Hudson).

Gardiner, T. (1991) *The York Institute and Masonic Jar*, unpublished.

Harrison, J. F. C. (1961) *Learning and Living 1790–1960: A Study in the History of the English Adult Education Movement* (London: Routledge and Kegan Paul).

Hird, D. (1985) *History of Spen Valley 1780–1980s* (Bradford).

Huddersfield and Holmfirth Examiner (1 August 1857).

Huddersfield Chronicle (18 June 1880).

Huddersfield Examiner (29 April 1882).

Huddersfield Female Education Institute (1858) *Annual Report*.

Huddersfield Female Education Institute (1875) *Annual Report*.

Maltby, J. (2001) 'The wife's administration of the earnings? Working-class women and savings in the mid-nineteenth century', *Continuity and Change* Vol. 26, No. 2, 1–31.

Moss, M., and Russell, I. (1994) *An Invaluable Treasure; a History of the Trustees Savings Bank* (London: Weidenfeld & Nicolson).

Muthesius, S. (1972) *The High Victorian Movement in Architecture 1850–1870* (London: Routledge and Kegan Paul).

Popple, J. (1991) *The Origins and Development of the Yorkshire Union of Mechanics' Institutes*, unpublished MA thesis (Nottingham: Nottingham University).

Rose, J. (2001) *The Intellectual Life of the British Working Classes* (Yale: Yale University).

Ross, D. M. (2002) 'Penny banks in Glasgow, 1850–1914', *Financial Review* Vol. 9, No. 1, 22–39.

Skipton New Mechanics' Institute and Science and Art Schools Building Committee (1894) *Statement*.

St John, A. J. (1858) *The Education of the People* (London: Chapman and Hall).

Stobart, J. (2004) 'Building an urban identity. Cultural space and civic boosterism in a 'new' industrial town: Burslem, 1761–1911', *Social History* Vol. 29, No. 4, 485–498.

Walker, M., and Haigh, B. (2009) '125th anniversary of the opening of the Ramsden Huddersfield Mechanics' Institute', public lecture, extracts in *Huddersfield Examiner* 6 July 2009.

The Worshipful Company of Clothworkers (1875–1914) *Trusts and General Superintendence Minutes* (Company archives: London).

Yorkshire Union of Mechanics' Institutes (1839) *First Annual Report*.

Yorkshire Union of Mechanics' Institutes (1845) *Seventh Annual Report*.

Yorkshire Union of Mechanics' Institutes (1849) *Eleventh Annual Report*.

Yorkshire Union of Mechanics' Institutes (1850) *Twelfth Annual Report*.

Yorkshire Union of Mechanics' Institutes (1852) *Fourteenth Annual Report*.

Yorkshire Union of Mechanics' Institutes (1853) *Fifteenth Annual Report*.

Yorkshire Union of Mechanics' Institutes (1857) *Nineteenth Annual Report*.

Yorkshire Union of Mechanics' Institutes (1858) *Twentieth Annual Report*.

Yorkshire Union of Mechanics' Institutes (1859) *Twenty-First Annual Report*.

Yorkshire Union of Mechanics' Institutes (1860) *Twenty-Second Annual Report*.

Yorkshire Union of Mechanics' Institutes (1862) *Twenty-Fourth Annual Report*.

Yorkshire Union of Mechanics' Institutes (1867) *Twenty-Ninth Annual Report*.

Yorkshire Union of Mechanics' Institutes (1869) *Thirty-First Annual Report*.

Yorkshire Union of Mechanics' Institutes (1873) *Thirty-Fifth Annual Report*.

Yorkshire Union of Mechanics' Institutes (1874) *Thirty-Sixth Annual Report*.

Yorkshire Union of Mechanics' Institutes (1875) *Thirty-Seventh Annual Report*.

Yorkshire Union of Mechanics' Institutes (1877) *Thirty-Ninth Annual Report*.

Yorkshire Union of Mechanics' Institutes (1879) *Forty-First Annual Report*.

Yorkshire Union of Mechanics' Institutes (1880) *Forty-Second Annual Report*.

Yorkshire Union of Mechanics' Institutes (1881) *Forty-Third Annual Report*.

Yorkshire Union of Mechanics' Institutes (1882) *Forty-Fourth Annual Report*.

Yorkshire Union of Mechanics' Institutes (1887) *Forty-Ninth Annual Report*.

Yorkshire Union of Mechanics' Institutes (1889) *Fifty-First Annual Report*.

Yorkshire Union of Mechanics' Institutes (1890) *Fifty-Second Annual Report*.

Yorkshire Union of Mechanics' Institutes (1902) *Sixty-Fourth Annual Report*.

8 Mechanics' Institute Libraries and their contribution to a public readership

Introduction

The majority of libraries established prior to the foundation of the Mechanics' Institute movement during the 1830s were for the emerging middling and upper classes. They were patronised in the main by the clergy, gentry and professional men (Kelly 1966). Books were expensive and those hired out through circulating libraries for one d. (one pre-decimal penny) 'were not always in the best improving taste' (Royle 1987: 249). With the emerging mechanics' institutes and other similar institutions, and subsequent levels of literacy increasing, there was a need for access to books and periodicals. The Sunday School Movement for both children and adults, founded in 1783, and the developments associated with Joseph Lancaster and Andrew Bell resulted in both Nonconformist and Church of England schools being established across the country. These and other developments put pressure on the need for public access to books, until now only available to those who could afford to be members of subscription libraries due to the cost of printing (Walker 2013). By the 1830s, technical developments in making 'machine-made paper and improvements in making printer's plates more efficient slashed the cost of large-scale print runs' (Allen 2010: 258–9). Thus, there was an increase in supply, which was within the expenditure of most libraries, if not individuals, as well as an increase in demand as literacy rates increased. There was also a significant increase in the number of newspaper titles, from 50 in 1753 to over 240 by 1833 (ibid.: 259).

Mechanics' institutes from their foundation always attempted to build up a stock of books for their membership. The addition of newspapers to libraries and reading rooms of mechanics' institutes supported additional sources of reading. They provided 'much incidental general education, and encouraged the habit of reading' (St Claire 2007: 267). By 1851, Kelly (1992) believed there were, in total, 655,000 volumes, an average of less than 1,000 per institute. The Liverpool Mechanics' Institute was an exception, with over 15,000 volumes. In contrast, a village institute might have just one shelf of books (ibid.: 175). In rural communities, institute reading rooms were often established first with classes being offered later. Kendal Mechanics' Institute in Westmorland, for example, had previously been founded as a working men's library, and similarly styled operative libraries continued well after 1850.

Mechanics' institutes had libraries, the larger ones having their own collections of books, newspapers and journals. Smaller institutes often borrowed books from their unions. In particular, the Union of Lancashire and Cheshire Institutes, the Northern Union and the Yorkshire Union provided a service whereby books were sent on loan to the smaller institutes, and the stock was exchanged on a fairly regular basis. With regard to the Yorkshire Union, Greenwood (1892: 56) observed that

> boxes with fifty or a hundred books are sent out periodically to mechanics' institutes and working men's clubs, and the books find their way into every part of that great county; the weaver, the ploughman, the collier, and the fisherman are all reached by that association.

The wooden boxes used for delivery and collection were designed in such a way that, when emptied of their books, they could be put on end with shelves, which were included, and used as bookcases. This arrangement was common practice in other parts of the world, including Australia.

In 1850, the 600 English institutes listed by Hudson (1851: 222–36) had between them almost 700,000 volumes and issued 1,820,000 books a year. The Yorkshire Union of Mechanics' Institutes had on average 900 books per institute (ibid.: 229–32). Frank Curzon, former President of Huddersfield Institute and the Agent for the Union, stated in 1898 that book boxes were still being circulated and 'two-hundred remote settlements were receiving fifty-volume shipments from the central stock of 40,000 volumes' (Altick 1957: 222).

Duckett (2003) highlights that some institute committees accepted members of the local community who had no wish to join classes or attend lectures but did want to make use of their libraries, paying lower fees for borrowing books. At the Hebden Bridge Institute in the Yorkshire Calder Valley, for example, non-members were charged one penny a visit to borrow books, hoping that at some stage they might take up full membership so as to attend lectures and classes. Other institutes, most notably Long Preston Institute in the Yorkshire Dales, established their own satellite libraries in surrounding villages under the administration of the Institute Committee.

Village circulating libraries

Circulating libraries can be traced back to the seventeenth century, but it was not until 1817 that Samuel Brown established rural libraries in East Lothian, which were supported by both philanthropists and borrowers. Later, many of these libraries became part of their local mechanics' institutes (Popple 1960).

In 1852, James Hole, Honorary Secretary of the Yorkshire Union, wrote a letter to the *Leeds Mercury* quoting the *Third Report of the Northern Union of Literary, Scientific and Mechanical Institutions*, which organisation had recently founded a Union itinerating library. He cited the successful example of the United Villages Perambulating Library, which had 300 volumes supplying nine Cumberland villages. Hole stated: 'small populations often had neither the inclination nor the means to organise an institution and therefore some external force ought to intervene to offer assistance' (ibid.: 134). He was indicating that the Yorkshire Union would be ideally suited to being that 'external force'. Hole and other Yorkshire Union Committee members had identified that such libraries were both cheap and simple to run and organise. Hole 'saw no difficulty in raising money for volumes and to set up a local reading room or library which would become self-supporting and perhaps develop into a centre for lectures and classes; in effect a small scale mechanics' institute' (ibid.: 137).

The Yorkshire Union had been set up with no specific objective to establish and manage village libraries, but it was able to do so through its general aims of supporting 'mental improvement'. It does indicate strongly that the Union was responding to a new initiative that would support the mechanics' institutes and encourage an increase in membership, underpinned with books to support attendees with their education. In order for a village library to be affiliated to the Yorkshire Union, there had to be at least 25 subscribers paying one penny a week or one shilling per quarter. The Union then selected 50 books, which were transported to the relevant settlements. Over several years, as finances allowed, the number of books on loan was increased (*Second to Twenty-Second Annual Reports of the Yorkshire Union* 1840–1860).

Within the Yorkshire Union there was the Bolton Abbey Group of libraries, which were founded in 1877. They were supported by the largest landowner in the Dales, the Duke of Devonshire, who provided rent-free land for the libraries in the villages of Barden, Halton East, Hazlewood and Bolton Bridge, with Bolton Abbey being the main centre for distribution (*Thirty-Ninth Annual Report of the Yorkshire Union* 1877: 105). Although there is no evidence in the reports sent to the Yorkshire Union that they were offering classes, they did provide the opportunity for the scattered communities to use them for reading and discussion.

In 1880, the Committee for the Bolton Abbey Group stated that 'each village depends upon the Yorkshire Union for their successive boxes of books and the circulation of volumes is most encouraging' (*Forty-Second Annual Report of the Yorkshire Union* 1880: 88). This cluster of libraries covered 10,000 acres with a total population of 700, offering both fiction and non-fiction books (*Forty-Seventh Annual Report of the Yorkshire Union* 1885: 97). The Yorkshire Union Village Libraries lasted for nearly 70 years, after which they became part of the West Riding of Yorkshire Library Service' (Duckett 2003).

The Northern Union of Mechanics' Institutes, covering Northumberland and Durham, had also introduced a similar itinerating library service to support smaller institutes that could not afford their own books. In 1856, a book, *The Northern Poetical Keepsake*, was published to raise funds for supporting the itinerating library (Stockdale, 1993: 223).

Literature and the Mechanics' Institute Libraries

Both the itinerating libraries stock and that permanently held by mechanics' institutes provided access to those who would otherwise not have been able to have the opportunity to read books. Stockdale (1993) suggests that most institute libraries in the Northern Union at least were probably catalogued in a similar way to the one at the Elswick Works Institute. In 1878, the library book stock was catalogued under 19 headings as listed in Table 8.1.

Mechanics' institutes prevented political or religious discussion through their rules of membership from their earliest beginnings. However, the Barnard Castle Mechanics' Institute, some 21 miles south of Durham, reported in 1897 that new books included Thomas Kirkup's *History of Socialism* and Benjamin Kidd's *Social Revolution*, indicating that institutes were not the hotbed of politics their founding fathers thought they might become. Religious books had also found their way onto the shelves (Stockdale 1993: 333). Most institutes also took daily and weekly newspapers as well as magazines and periodicals so, had they continued to ban religious and political books, members still would have had the opportunity to keep up to date with debates through newspapers. West Hartlepool Literary and Mechanics' Institute took '22 daily newspapers, 21 weekly papers and 22 of the best magazines' (*West Hartlepool Literary and Mechanics' Institute Annual Report* 1893). Obviously, smaller numbers of newspapers and periodicals were taken by minor institutes where membership was much lower.

The Mechanics' Institute Village Library itinerary schemes supporting the Northern Union seem to have delivered 20 books in each box to their libraries. Table 8.2 is a sample indicating the kinds of books read by members between 1888 and 1894, and publications would have been similar for libraries across the whole of the British Isles.

Table 8.1 Elswick Works Institute Library catalogue

Class	Subject
A	History
B	Biography
C	Astronomy, geography and navigation
D	Chemistry
E	Botany
F	Geology and mineralogy
G	Designing, perspective, etc.
H	Science and art
J	Mechanics and engineering
K	Architectural building
L	Miscellaneous
M	Mathematical and educational
N	Voyages, travel and narratives
O	Moral and religious
R	Fiction
S	Poetry
T	Magazines and reviews
U	Reports
W	Reference

Source: Stockdale 1993: 332–3.

Table 8.2 The Mechanics' Institute Village Library books, sample of the Northern
Union 1888–1894

Disraeli *Alroy* and *Sybil*	*Ancient Spanish Ballads*	Disraeli *Vivian Grey*	Croker *History of the Guillotine*
Bunyan *Puritan Divines*	Wittich *Physical Geography*	*Philosophy of History*	Dickens *American Notes*
Barrow *Tour of the Continent*	Trollope *Vicar of Bulhamton*	Dilke *Greater Britain*	*Pompeii*
Smiles *Life of Telford*	Dickens *Martin Chuzzlewit*	Shakespeare *The Works of ...*	McCulloch *London in 1850–1851*
Proctor *Light Science*	*Life of Louis XIV* (two volumes)	Gostick *German Literature*	*The Life of Hutchinson*
Smiles *Self-Help*	Larmartine *French Revolution*	*Vegetable Substances: Materials of Manufactures*	*Early Years of the Prince Consort*
Brougham *Lives of the Philosophers*	Reid *On Chemistry*	Hudson *Adult Education*	*Border Tales*
Dickens *David Copperfield*	Trollope *Orley Farm*	Whewell *Astronomy*	*British Costume*

Source: Stockdale 1993: 386–9.

While these are a sample, it is interesting to note that there were few books
relating to academic subjects, the exception being Reid's book, *On Chemistry*.
On the other hand, novels are well represented, particularly Trollope, Disraeli
and Dickens; the latter spent a great deal of time travelling across the country
presenting readings of his works at various mechanics' institutes. It is also interesting to note the publications on the French Revolution and the history of the
guillotine, deemed non-controversial some 99 years after the events took place
in Paris.

At Huddersfield, the *Catalogue of Foreign Books* for the Technical School and
Mechanics' Institute (circa 1883: 1–5) included, under the heading of French
books, *Voltaire, The History of France, Geography of France, The Speech of Louise
XIV, Les Miserables* and other works of Victor Hugo, *The French Revolution* and
Napoleon Bonaparte. The German collection included works on *Buildings, Kaiser
Leopold, Lord Byron* and *Political Buildings*. There were also Spanish and Italian
books; in the case of the latter, most were on operas and travel. Many institutes
offered French, and in the case of Huddersfield it is known that it was offered
along with German, both of which were popular for those members involved in
the export trades associated with textiles.

The *General Catalogue* of the Technical School and Mechanics' Institute for
1885 to 1886 was divided into two parts, fiction and non-fiction. Fiction included
Jane Austen's *Mansfield Park, Pride and Prejudice* and *Sense and Sensibility*, and
Anne Bronte's *Agnes Grey* and *The Tenant of Wildfell Hall*. There were several
copies of Charlotte Bronte's *Jane Eyre, The Professor, Shirley* and *Villette* as well as
Emily Bronte's *Wuthering Heights*. There were also the works of Lewis Carroll,
Alice's Adventures in Wonderland and *Through the Looking Glass*, 15 works of

Charles Dickens and seven, including *Sybil*, by Benjamin Disraeli. There were also books on poetry and drama.

Non-fiction books included various publications on trade, manufacturing, dyeing, spinning and weaving. Other themes included works on chemistry, natural philosophy including philosophy, physics, mechanics, engineering, steam and the steam engine, heat and general science. Other subjects included politics, economics, law, currency, geology, agriculture, bookkeeping, shorthand, architecture, fine art and drawing. Finally, there were biographies, books on travel, history and local interest. There were eight volumes by Thomas Twining, titled *Technical Training*, published in 1874.

Some institutes, such as the one at Hebden Bridge in Yorkshire, allowed non-members to use the library, provided they paid a very low charge for the privilege, and this expanded the readership. Moran (2006) has identified that surviving catalogues of institute libraries highlight that liberal and literacy publications were accessible, along with Victorian literature, to the membership. Rauch (2001: 135) highlights that at Keighley Mechanics' Institute the Rev. Patrick Bronte and his family regularly made use of its library, Bronte himself organising and delivering lectures at the Institute. Charlotte Bronte, his literary daughter, was known to visit the Mechanics' Institute Library at Keighley and 'there is little question that she admired the principle of making useful knowledge available to the working classes'.

The Public Libraries Act of 1850

Traditionally, and certainly prior to the 1850s, a committee member of the local mechanics' institute would often manage the library. However, as Stockdale (1993) highlights from the 1850s, many larger institutes employed a librarian, the first time possibly a permanent position of such a post was offered. At Berwick-upon-Tweed, on the English–Scottish border, an advert was locally put out in 1853. It stated that the successful candidate would 'reside upon the premises, [and] attend in the library for the purpose of giving out and receiving books'. It further stated that

> for his services ... shall have one room, coals and gas free of charge ... five pounds per annum ... to be paid quarterly ... and shall also receive one shilling in the pound for all money collected out of the Institute (*Berwick-upon-Tweed Mechanics' Institute Minutes* 16 May 1853).

At other institutes, notably Middlesbrough, the responsibilities of librarianship also involved more general duties as a porter, which included collecting member subscriptions for the next quarter and canvassing the local population for members (Stockdale 1993: 221). At Gateshead Mechanics' Institute, the committee made a joint appointment of librarian, given to William Stockdale, and keeper of the rooms, given to his wife. Their joint salary was £18 per year.

The introduction of The Public Libraries Act of 1850 permitted municipal authorities in England and Wales to provide library premises on the understanding that two-thirds of the rate-payers agreed to support them. Books were an additional cost. The Act was extended in 1853 to include Scotland, and in 1854 Ireland. A further Act, passed in 1855, provided the implementation of a one d. (one pre-decimal penny) levy and permission for further expenditure on buildings, books, newspapers, maps and specimens of art and science (Kelly 1957: 177). This caused much concern amongst mechanics' institute committees. Their libraries had often been the central focal point, a lifeline, and provided access to a wider population in exchange for a fee. As well as supporting mental improvement for the working classes, they often encouraged the local populace to then attend public lectures and sign up for classes. There, museum collections and works of art were part of this service. However, the Acts at first did not have the impact on the Mechanics' Institute Movement to the extent many committees feared. In 1869 there were only 35 English, Scottish and Welsh local authority libraries, and only by 1900 had their numbers increased to 266 (Hamilton 2003: 107–8). Indeed, Kelly (1966) states that it was assumed that the 1850 Act 'was a magic wand that brought a library service into being overnight'. He highlights that the powers of the Act were exceedingly limited and inadequate and were 'permissive not compulsory' (ibid.: 238). Thus, it was to be a slow change, and institute libraries continued to be important until the end of the nineteenth century.

The mechanics' institutes were not in a position to object to free public libraries, as they were state-funded. Both had an aim to support the mental improvement of the working classes. Middlesbrough Mechanics' Institute Library is one example where it was taken over by the town under the 1850 Act, but not until 1873. Understandably, the Committee opposed the decision, but members did accept that there would be more choice of books for residents in the town. It agreed to hand over the reading room and librarian's house in support of the town's Free Library. The Committee's fears were well-founded, as 'all inhabitants of Middlesboro' being free to attend the reading room, and get books from the free library, we have no members of the Institute save those who attend classes' (*Thirty-Fifth Annual Report of the Yorkshire Union of Mechanics' Institutes* 1873: 102). The Committee did, however, accept that the Free Library, with a circulation of 33,752 books and the addition of a further 932 volumes in its first year, 'was far beyond what the Mechanics' Institute could have achieved' (ibid.: 102). A few miles away, a Free Library was also established at the Darlington Institute in 1887 where the Committee had delayed the decision as it thought it 'would be injurious to the existing subscription library and the library of the mechanics' institute' (Kelly 1966: 62) Following the foundation of a Free Library, however, the Institute was able to continue to offer classes and keep its library for subject–based classes (*Forty-Ninth Annual Report of the Yorkshire Union of Mechanics' Institutes* 1887). At Newcastle-upon-Tyne, the Mechanics' Institute was transferred to the Free Library Committee of the Corporation in 1878. Under the agreement, classes and lectures were to 'be continued and extended ... under the title of the Newcastle Mechanics' Institute Free Library' (Stockdale 1993: 295). However, it was not always a successful partnership between the Free Library and the local institute. At the West Hartlepool Institute, for example, in 1891, there were substantial disagreements between the Institute and the Free Library, with both parties standing their ground. In 1893, the Institute Committee wrote to the Corporation stating that they had no further points to add to their argument for not being taken over. This was the last correspondence between the two and the result was that the Institute continued to offer classes and lectures, supported by its library, and a separate Free Library was built in the town some years later (Stockdale 1993: 296–7).

The Institute Movement was, however, losing ground. By 1868, for example, Manchester Public Library had received 2000 volumes donated by the Miles Plating Mechanics' Institute Library (Kelly 1966: 74). There was obviously little point in having a fee-paying library, which subsequently saw a decline in its usage, even where classes might continue to be well attended. Some institutes were not even fortunate enough to keep their book stock for classes as the result of the 1850 Act. Leicester Mechanics' Institute ceased to exist in 1871 after all its books were transferred to the town's recently opened Free Library (Lott 1935). Other institutes did continue until the end of the century and beyond, with their libraries still intact, including those at Bradford, Burnley and Huddersfield (Altick 1957). Bradford still has its Mechanics Institute Library, where members pay an annual subscription to use its facilities.

Despite the impact of the 1850 Libraries Act, mechanics' institutes did make a substantial contribution to free facilities for the working-class readership. Few of the public libraries had their own buildings to begin with and so it was quite common for a mechanics' institute to 'surrender its premises [or at least its library], either for free or at a modest charge on the understanding that the service it had formerly provided for its members would now become public responsibility' (Kelly 1966: 66). This at least meant members of the public would be attending the institutes while using the Free Library. While in the library, they would have been made aware of public lectures, classes and entertainments.

Some mechanics' institute buildings became public libraries. Sheffield was the first public library opened in Yorkshire in 1856. With no building of its own, it rented a room in the Institute. By 1860, the Free Library was renting additional rooms and did so until the Institute closed at the end of the century when it became the town's public library (ibid.: 52). In 1870, newly opened public libraries and newsrooms at Hunslet and New Wortley, both suburbs of Leeds, were accommodated in their former institutes, as well as inheriting their institutes' books. The Pendleton Mechanics' Institute in the suburb of Manchester reopened in 1877 as a branch public library attached to Salford Central Library (ibid.: 49–52) There are examples of this arrangement in all parts of the country, such as at Bolton, Newcastle-upon-Tyne, Norwich, Nottingham and Sunderland (ibid.: 53–60).

The Carnegie Libraries and the Mechanics' Institute Movement

What was becoming apparent to institute committees was that public free libraries were also supported by benefactors, notably Andrew Carnegie, which saw an increase in library building between 1897 and 1914 (Royle 1988: 250).[1] Carnegie, the Scottish-born American oil and steel magnate and philanthropist, provided funds to support the building of libraries in several towns throughout Britain as well as overseas. His criterion was that funds would be available to towns that were committed to purchasing books on behalf of their townsfolk, a decision he only made if his Trust was sure libraries would provide the book stock and that they would be well used. Thus, mechanics' institutes were ideally placed to support this as they had encouraged the local population to make good use of their libraries. It is no coincidence that many mechanics' institute towns were the same ones that had a Carnegie Library, the latter providing free access to all. With the passing of the Public Libraries Act in 1850, the mechanics' institute became the repository while funds were found for a purpose-built library where a Carnegie Library was not planned.

At Skipton in the Yorkshire Dales, for example, Carnegie provided £3000 in 1903 to build and furnish a local library which shared the same building as the Institute, located in the market square, and which is part of Craven College of Further Education premises today (Gibbon 1958: 6). Carnegie also donated £10,000 to establish a public library in Keighley, following a meeting with his friend Sir Swire Smith, who was President of the Keighley Mechanics' Institute. Smith convinced Carnegie that, even though the town's mechanics' institute was being well utilised, there was a desperate need for a public library (*One Hundred Years of Keighley Library* 2004: 5.) Under the Free Library Act, the Borough of Keighley was able to give land for the building, which was located directly across the road from the Institute (ibid.). When the former mechanics' institute at Keighley became a local authority college, it restocked its library with books relevant to its courses and donated the rest to the Carnegie Library.

Conclusion

Mechanics' Institutes Committees had always seen the importance of their libraries as contributing to the education of their membership. Unions provided Village Library schemes that provided the loan of books to the smaller centres, which in any case often had to rely on restricted accommodation. The larger institutes were able to stock their own books, in some cases provided as gifts and also through the income raised from membership fees. Both fiction and non-fiction books were available, and towards the end of the nineteenth century political and religious publications were often available for members. Libraries were available for males and females as well as families. Thus, libraries were an important part of 'mental improvement'. With the passing of the 1850 Library Act, and the Free Libraries that were eventually established across the country, mechanics' institutes either provided space for the town library or found itself in competition with it. By the end of the nineteenth century, as the mechanics' institutes began to decline and be replaced by technical schools and colleges, their books were often transferred to these institutions or the Free Library. Crucially, the mechanics' institute libraries had provided a foundation on which free public libraries became established.

Note

1 Andrew Carnegie was born in 1837. His father was a Chartist, and he started as a bobbin boy in a cotton factory in America. Later he became a telegraph clerk, a railway superintendent, and then worked in the oil industry, finally becoming an oil and steel magnate with a fortune of £10 million, which was used for establishing and supporting libraries both in America and the UK. In total, Carnegie established 2,811 libraries in Scotland, England and America (Keighley News 2004: 5).

References

Berwick-upon-Tweed Mechanics' Institute (16 May 1853) *Minutes.*

Allen, D. (2010) *Commonplace Books and Reading in Georgian England* (Cambridge: Cambridge University Press).

Altick, R. D. (1957) *The English Common Reader: A Social History of the Mass Reading Public, 1800–1900* (Chicago: University of Chicago Press).

Duckett, B. (November 2003) 'From village hall to global village: Community libraries in England's largest county', *Library History* Vol. 19.

Gibbon, A. M. (1958) *Skipton Mechanics' Institute* (Skipton: The Craven Herald).

Greenwood, T. (1892) *Sunday-School and Village Libraries* (London: J. Clarke and Co.).

Hamilton, W. R. R. (2003) *Development and Change in Municipal Ideology and Practice: The Role and Expansion of Urban District Councils to 1940 with Special Reference to County Durham*, unpublished MPhil thesis (Huddersfield: University of Huddersfield).

Huddersfield Technical School and Mechanics Institute (circa 1883) *Catalogue of Foreign Books* (Huddersfield: Alfred Jubb).

Huddersfield Technical School and Mechanics' Institute (1885–1886) *General Catalogue* (Huddersfield: Alfred Jubb).

Hudson, J. W. (1851) *The History of Adult Education in which is comprised a Full and Complete History of the Mechanics' and Literacy Institutions* (Reprint. London: Woburn, 1969).

Keighley News (2004) (author unknown) 'One hundred years of Keighley Library 1904–2004' (Keighley: Keighley News).

Kelly, T. (1957) *George Birkbeck: Pioneer of Adult Education* (Liverpool: Liverpool University Press).

Kelly, T. (1966) *Early Public Libraries: A History of Public Libraries in Great Britain Before 1830* (London: The Library Association).

Kelly, T. (1992) *A History of Adult Education in Great Britain from the Middle Ages to the Twentieth Century* (Liverpool: Liverpool University Press).

Lott, F. B. (1935) *The Story of the Leicester Mechanics' Institute 1833–1871* (Leicester: Thornley).

Moran, M. (2006) *Victorian Literature and Culture* (London: Continuum International Publishing Group).

Popple, J. (1960) *The Origin and Development of the Yorkshire Union of Mechanics' Institutes*, unpublished MA dissertation (Sheffield: University of Sheffield).

Rauch, A. (2001) *Useful Knowledge: The Victorians, Morality, and the March of Intellect* (Durham: Duke University Press).

Royle, E. (1988) *Modern Britain: A Social History 1750–1985* (London: Edward Arnold).

St Claire, W. (2007). *The Reading Nation in the Romantic Period* (Cambridge: Cambridge University Press).

Stockdale, C. (1993) *Mechanics' Institutes in Northumberland and Durham 1824–1902*, upublished PhD thesis (Durham: Durham University).

Walker, M. (2013) 'For the last many years in England everybody has been educating the people, but they have forgotten to find them any books: The Mechanics' Institutes Library Movement and its contribution to working-class adult education during the nineteenth century', *Library and Information History* Vol. 29, No. 4, 272–286.

West Hartlepool Literary and Mechanics' Institute (1893) *Annual Report.*

Yorkshire Union of Mechanics' Institutes (1873)*Thirty-Fifth Annual Report.*

Yorkshire Union of Mechanics' Institutes (1877) *Thirty-Ninth Annual Report.*

Yorkshire Union of Mechanics' Institutes (1880) *Forty-Second Annual Report.*

Yorkshire Union of Mechanics' Institutes (1885) *Forty-Seventh Annual Report.*

Yorkshire Union of Mechanics' Institutes (1887) *Forty-Ninth Annual Report.*

9 Analysis of three clusters of mechanics' institutes and their impact on their local communities

Introduction

As previously mentioned in Chapter 1, Mathias, among others, highlighted the increasingly close links between history and other disciplines, such as education, as did Sanderson (2007), who referred to them as neighbours. Marsden (1977) researched the relationship between historical geography and the history of education and this approach supports the theme of this chapter. As a result of research carried out with regard to the Yorkshire Union, three clusters have been selected for further research in order to analyse the impact institutes had on their localities.

The North East Cluster

The North East was chosen for its rapid growth in the number of institutes, many of which were specifically mining institutes established and funded by the Joseph Pease Mining Company, which had invested heavily in both iron ore and coal mining exploration. The Pease Company also established separate day schools for children in their mining communities. There were other mechanics' institutes in County Durham, but they were members of the Northumberland and Durham Union. It therefore seems likely that it was a personal choice of the Pease and Partners Mining Company to be associated with the Yorkshire Union despite its headquarters at Leeds being 90 miles south of Stanley, the farthest northern Yorkshire Union Institute, near Durham. This may have been due to the personal preference of the Pease family, several members of whom were on the committees of established institutes, and who were also involved with founding their own in both County Durham, in coal mining communities, and in the ironstone communities of the North Riding of Yorkshire. The Pease Family purchased land by the River Tees and built an industrial settlement that became Middlesbrough. It was the discovery of iron ore in the Cleveland Hills in 1850 that contributed much to the expansion of Middlesbrough. As Kirby (1984: 32) points out, 'what happened to the economy of Teesside after 1850 hardly counted as a revival in prosperity; it was regional economic growth of an altogether different pace and magnitude'.

The number of mechanics' institutes established in this cluster, many in the newly developing mining communities, reflects the industrial activity taking place in Country Durham and North Yorkshire (see Figure 9.1 and Table 9.1).

The early institutes were general ones, such as those at Darlington, Hartlepool, Middlesbrough, Redcar and Stockton, and which were founded during the period 1825 to 1840. During the 1850s, new institutes were established such as those at Darlington (railway), Crook (coal) and Eston (ironstone). Further developments took place from the 1860s and several institutes reflect the advanced industrial developments along the River Tees, such as the Cleveland Ironworks and Middlesbrough South Bank Institutes. Those institutes established during the 1870s and 1880s were all associated with mining, such as those at Peases West, Tanfield and Stanley in County Durham, and Upleatham, Liverton Mines and Brotton in the Cleveland foothills, as well as several others not specific, including Skinningrove, along the coast of North Yorkshire. Several Durham institutes in this cluster went on to become Science and Art Schools. It was not just the general institutes at Darlington and Stockton that became science and art schools but also several former institutes in the mining communities such as those at Esh and Crook (*Annual Reports of the Yorkshire Union of Mechanics' Institutes,* from 1877). The Pease Company also introduced a circulating library that supported their institutes, a similar arrangement to that established by the Yorkshire Union. There were other mechanics' institutes that were not members

Table 9.1 The North East Cluster of Yorkshire Union Mechanics' Institutes relating names to numerical locations in Figure 9.1

1	*Stanley	14	Spennymoor	27	South Bank, Middlesbrough	40	Loftus
2	Chester-le-Street	15	Howden-le-Wear	28	Coatham	41	North Skelton
3	Lanchester	16	Ferry Hill	29	Redcar	42	Seklton
4	Pittington (not YUMI)	17	Hartlepool	30	Marske-by-the-Sea	43	Guisborough
5	Durham (not YUMI)	18	Etherley	31	New Marske	44	Lingdale
6	Sherburn	19	Sedgefield (not YUMI)	32	Kirkleatham	45	Liverton
7	Esh	20	Shildon	33	Eston and Normanby	46	Moorholm
8	Waterhouses	21	Stillington	34	Middlesbrough	47	Great Ayton
9	Sunnyside	22	Norton	35	Upleatham	48	Stokesley
10	Castle Eden	23	Port Clarence	36	Saltburn	49	Seamer
11	Hutton Henry	24	Stockton-on-Tees	37	Brotton	50	Forcett
12	Billy Row	25	Darlington	38	Skinningrove	51	Eppleby
13	Crook and Peases West	26	Denton	39	Staithes	52	Yarm

* Stanley Mechanics' Institute, located ten miles south-west of Newcastle, was the most northern in the Yorkshire Union.

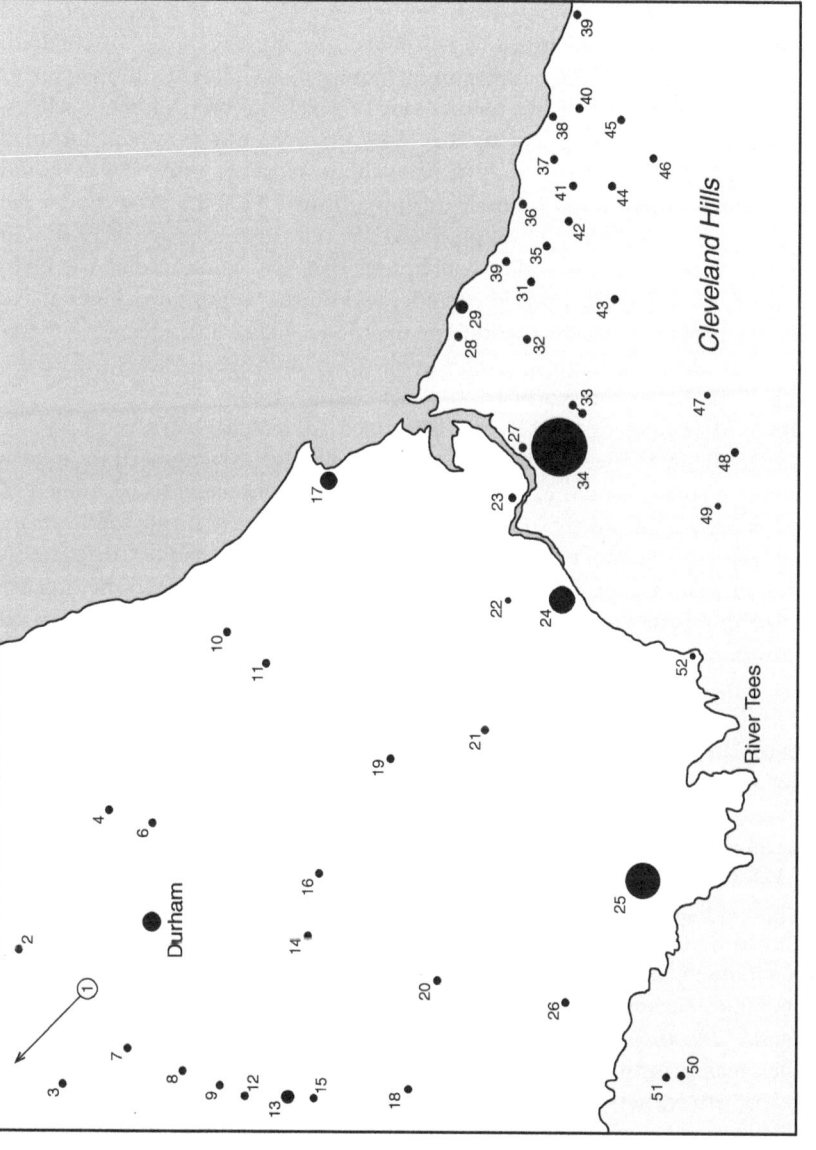

Figure 9.1 The North East Cluster of the Yorkshire Union of Mechanics' Institutes.

of the Yorkshire Union but were established in County Durham. These included Sunderland, Bishop Auckland, Gateshead, Leadgate, and Consett, and at least a further 12 miners' institutes were established from 1877 onwards (Stockdale 1993: 371–4).

The coal mining institutes offered elementary education to the children of miners, in areas where schools were not yet established. As a result of Lord Ashley's 'Ten Hours Bill' in 1833 to reduce the working hours of children, Parliament set up a Royal Commission of Inquiry into Children's Employment. The Commission produced two reports, one of which was on Children in Mines, published in May 1842. It led to the 1842 Coal Mines Act, which included prohibiting the employment of all female labour and boys under 10 years old from working underground (www.legislation.gov.uk 1842). In 1864, under the Inspection of Mines Report, it was proposed that boys under the age of 12 should not work in the mines unless they could prove that they could read and write in support of reading instructions in collieries and on the working machines above and below ground (www.legislation.gov.uk 1864). These institutes were therefore in a strong position to offer education to children several years before the passing of the Education Act of 1870.

Mining developments also had an impact on adult miners. With technological advancements in the mining industry, there was the growing need for miners to be able to read and write, as there was a necessity for them to be trained to operate machinery correctly and safely. In a newspaper article entitled *Educate the Miners* a strong argument was put forward that miners who had no basic education should not be allowed to go down the pits as operators, due to the expense of replacing misused machinery. The Malicious Damage Act of 1861 threatened the prosecution of miners who damaged mining machinery, even if unintentionally, as a result of not receiving the required training (www.legislation.gov.uk 1861). The mining institutes were therefore able to offer both elementary and mining-related education.

Not surprisingly, these mining institutes in both County Durham and North East Yorkshire had smaller memberships than those in the industrialising town. For example, Castle Eden Colliery had 165 members in 1880, Crook Colliery had 156, Esh Colliery 230, Marske Institute 100, Peases West Colliery 400, Redcar Institute 218 and Skinningove colliery 360. This compares with Darlington, which had 554 but was a much larger town. Thus, per head of population, the small mining institutes were supporting more of their population with elementary and mining education. (*Forty-Second Annual Report for the Yorkshire Union of Mechanics' Institutes* 1880).

The smaller institutes tended to have steady membership throughout the period of study, serving their communities well regarding adult education. Decline in membership overall came late in the nineteenth century and seems to have been due to the development of technical and adult education in the larger towns of Darlington, Durham, Hartlepool, Middlesbrough and Stockton, all of which Institutes went on to become technical colleges. Some of the mining institutes became reading rooms or outreach centres.

The Dales and Pennines Cluster

What made the Dales and Pennines institute cluster unique was that many of them were in rural or semi-rural areas and yet were able to continue despite the challenges they faced in relation to rural depopulation and the impact of textile trade depressions on their membership (see Figure 9.2 and Table 9.2).

The first generation of mechanics' institutes, established between 1824 and 1850 in this cluster, included Keighley (1825), Skipton (1825), Bingley (1825) and Otley (1839). Skipton Mechanics' Institute was one of several that closed and reopened when economic and membership conditions allowed. However, after reopening in 1848 and again in 1866, it continued until finally becoming a further education college in the twentieth century (*Tenth Annual Report of the Yorkshire Union of Mechanics' Institutes* 1848; *Twenty-Eighth Annual Report of the Yorkshire Union of Mechanics' Institutes* 1866). There were further developments from the 1840s until the late 1870s in the Pennines and across the Yorkshire Dales. In the villages, they were often small institutes, commonly a small reading room that could be adapted into a classroom. These included Broughton with Elslack, near Skipton, and Coniston Kinsey and Haworth and Sedbergh near Settle. The mechanics' institutes in this cluster provided education

Table 9.2 The Dales and Pennines Cluster of Yorkshire Union of Mechanics' Institutes relating names to numerical locations in Figure 9.2

* Clitheroe	10 Tosside	28 Barnoldswick	46 Harden
* Colne	11 Halton West	29 Earby	47 Cottingley
* Padiham	12 Gargrave	30 Salterforth	48 Saltaire
* Burnley	13 Embsay	31 Lothersdale	49 Shipley
* Accrington	14 Halton East	32 Cononley	50 Wilsden
* Bacup	15 Hazlewood	33 Kildwick	51 Cullingworth
* Rawtenstall	16 Bolton Bridge	34 Silsden	52 Haworth and Stanbury
* Mankinholes	17 Skipton	35 Ilkley	53 Oxenhope
* Heptonstall	18 Broughton	36 Burley-in-Wharfedale	54 Denholme
1 Settle	19 Slaidburn	37 Otley	55 Denholme Clough
2 Grassington	20 Newton	38 Cowling	56 Lydgate
3 Burnstall	21 West Marton	39 Keighley	57 Todmorden
4 Airton	22 Addingham	40 Laycock	58 Eastwood
5 Long Preston	23 Carleton	41 Thawaites Brow	59 Hebden Bridge
6 Rylstone	24 Elslack	42 Eldwick	60 Mytholmroyd
7 Hellifield	25 Gisburn	43 Baildon	61 Wainstalls
8 Bell Busk	26 Bolton-by-Bowland	44 Oakworth	62 Luddenden Foot
9 Barden Scale	27 Thornton-in-Craven	45 Bingley	63 Cragg Vale

* Not members of YUMI, but members of the Lancashire and Cheshire Union of Mechanics' Institutes.

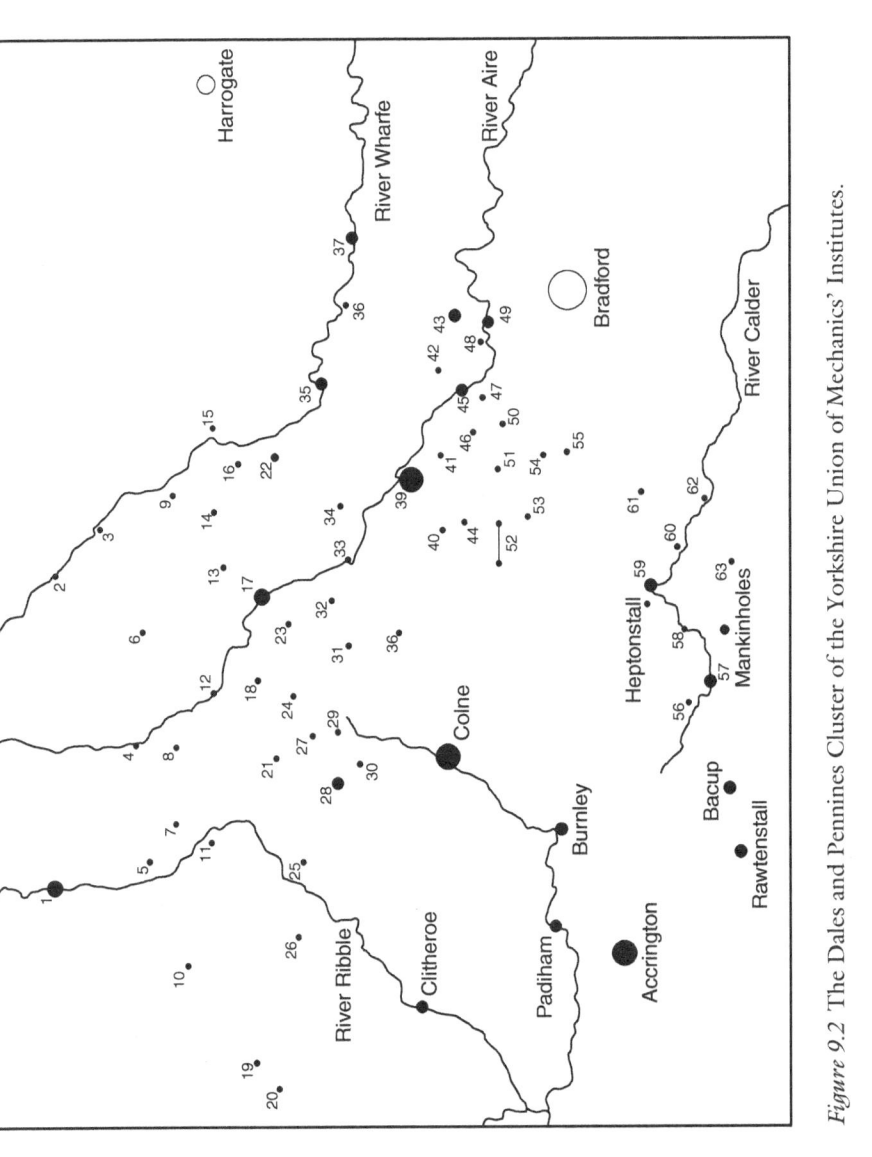

Figure 9.2 The Dales and Pennines Cluster of the Yorkshire Union of Mechanics' Institutes.

for those living in the small valleys and isolated settlements and small rural settlements. Often these agricultural communities relied on textiles for a living, with local rivers providing water power for the newly developing textile mills.

The mechanics' institutes benefited from the growth and needs of the textile industry, and courses were offered that were relevant to the needs of the employees both in the Pennines and Dales. At Oxenhope, for example, the Institute had the loan of a building, the rent of which was paid for by 'a few gentlemen [who were] heartily supported by the manufacturers and others' who funded the venture and obviously identified the potential regarding education and training in relation to their workforce (*Forty-First Annual Report of the Yorkshire Union of Mechanics's Institutes* 1879: 98.). Several went on to become art, technical and evening schools, or else libraries, that were still in use in 1910. The science and art schools in the Yorkshire Dales provide evidence that many of these institutes continued to be successful into the early years of the twentieth century, and several of them became further education colleges, including those at Saltaire (Shipley College), Skipton (Craven College), and Keighley (Leeds City College Campus).

While the textile industry supported institute development, it did mean that institutes were vulnerable during the trade depressions, with subsequent depopulation. The impact was felt by mechanics' institutes in the industrialising towns, but particularly by those in the rural communities. This did mean institutes often noted in their reports a fall in membership, and sometimes they closed. Addingham is one example. A rural community located between Skipton and Ilkley in the Wharfe Valley, the Institute was opened in 1845. However, membership had declined by 1851 to 41 as people moved out of the village in search of work,

> owing to the stoppage of the large cotton mills connected with this place, which has had a ruinous effect on the village at large. The number of houses in the village is somewhere about 500, of which not less than 150 are unoccupied and there does not appear to be the slightest chance of any improvement taking place soon (*Thirteenth Annual Report of the Yorkshire Union of Mechanics' Institutes* 1851: 26).

The two cotton mills stopped producing cotton and worsted in 1850, with the result that the population of Addingham and surrounding areas fell from 2200 to 1559. Families who stayed in the village were either unemployed or struggled on in part-time work such as agriculture (*Fourteenth Annual Report of the Yorkshire Union of Mechanics' Institutes* 1852: 39). The Mechanics' Institute did close but then reopened and indeed expanded to such an extent that additional accommodation had to be found by the 1870s. It only finally closed in 1930.

Settle, a small market town, was also affected by the trade depression and in 1860 it 'was pleased to report that there had been a vast improvement in the state of the Institute'. It had only been 12 months earlier that it had been 'seriously thinking of closing its doors due to rural de-population'. A female class was established, which seemed to have saved it from closing, and it went on to offer Society of Arts examinations, which was an attraction for members. These events indicate that committees were adapting to trade cycles by offering wider relevant curricula to a broader membership, in the hope of staying in business (*Twenty-Second Annual Report of the Yorkshire Union of Mechanics' Institutes* 1860: 119).

This was also true at Todmorden, where the Committee reported in 1862 that membership was in decline owing 'to the very depressed state of the staple trade of the district ... one of those places so severely afflicted by the dire disease, the cotton famine' (*Twenty-Fourth Annual Report of the Yorkshire Union of Mechanics' Institutes* 1862: 132). Due to the trade depression, boys whose families were unable to pay the membership were admitted for free, as were the unemployed, thus maintaining loyalty during the period of depression in the hope that they would pay their fees once there was an improvement in the economy and they were in employment once more (*Twenty-Fourth Annual Report of the Yorkshire Union of Mechanics' Institutes* 1862).

During the following year, the Todmorden Committee reported that it was more optimistic than previously, and that 'on the whole they [the Committee] consider the undertaking to be in a healthy and satisfactory state. The Institute was 'doing a considerable amount of good amongst the labouring population' (*Twenty-Fifth Annual Report of the Yorkshire Union of Mechanics' Institutes* 1863: 138). The town was still in a depressed state and had been 'one of those places severely afflicted by the dire disease of the cotton famine'. Nevertheless, there had been 16 lectures and six readings at the Institute, 'filling the room on all occasions' as well as clearing some debt (*Twenty-Fifth Annual Report of the Yorkshire Union of Mechanics' Institutes* 1863).

Four miles down the Calder Valley from Todmorden, the Committee at Hebden Bridge Institute reported in 1863 that, 'although the general prevalence of distress has severely tested the Institution, it continues to fulfil its useful office' in offering relevant subjects (ibid.: 100). In 1865, however, the Institute was struggling due to 'the depression of trade resulting from the continuance of the war in America'. (*Twenty-Fifth Annual Report of the Yorkshire Union of Mechanics' Institutes* 1865). Although Yorkshire is associated with woollen textiles, there were hundreds of cotton mills in the country, from Sheffield in the south to Sedbergh in the north. In the Todmorden and Hebden Bridge areas of the Calder Valley alone, there were over 40 mills that were producing cotton textiles at one time or another (Ingle 1997).

A decade later, mechanics' institutes in the cluster were feeling the effects of further economic decline. At Grassington in 1876, for example, the Institute Committee reported that 'the work is reviving here' following the depression in trade. An additional room was being built for evening classes and although many of the young people had moved away to the 'more populous places', it was confident that new members would be found (*Thirty-Eighth Annual Report of the Yorkshire Union of Mechanics' Institutes* 1876: 148). Otley Institute was also susceptible to the trade depressions. Nevertheless, it provided the facilities of a library, reading room, public lectures and education to both male and female members (ibid.: 189). The Dales and Pennines institutes were seemingly affected by trade depressions over two decades or so. However, they either reopened when trade was good, or adapted ways of encouraging attendance where they could through offering free access to the libraries, hoping that when times were better men and women would see the advantage of paying fees for the services offered. Secondly, many of the rural institutes continued to operate until almost the end of the nineteenth century, when most became public libraries and easier travel meant that both children and adults could attend the technical schools such as those at Skipton, Keighley, Halifax and Todmorden.

The Huddersfield and District Cluster

A cluster around Huddersfield was chosen to examine to the extent to which local developments were influenced or restricted by having a larger successful mechanics' institute nearby. There were 43 mechanics' institutes established within a ten-mile radius of Huddersfield between 1830 and 1891 and, unlike those in the North East and Dales and Pennines, the institutes in this cluster were closely located to each other, in several cases only two or three miles apart.

Evidence from the Committee Reports indicates that the institutes in and around Huddersfield benefited from their close geographic locations, supporting each other with the skills and expertise of teachers who were willing to work in more than one institute (*Thirty-Sixth Annual Report of the Yorkshire Union of Mechanics' Insitutes* 1874: 158) (See Figure 9.3 and Table 9.3).

Competition, though, was a problem at some of the institutes. While staffing supporting local institutes was a positive result of being close by, there was the potential problem of offering similar courses. The Committee at Dogley Lane Institute, some two miles south of Huddersfield, for example, noted that membership had increased by over 100 in 1847 and they saw this as quite an achievement as there was 'another institution about a mile and half distant, which has rapidly increased in its numbers; showing fully the anxiety that prevails amongst the working classes in this neighbourhood for education' (*Ninth Annual Report of the Yorkshire Union of Mechanics' Institutes* 1847: 35). The other 'institution' referred to was probably Almondbury. This suggests that institutes in the Huddersfield area were not under serious threat from competition as there were so many working-class adults anxious to have the opportunity to attend. Indeed, the Committee at Dogley Lane was particularly grateful to the Huddersfield Mechanics' Institute for allowing some of their members to attend its elocution classes, as, presumably, Dogley Lane had not enough students to offer additional subjects other than elementary ones (ibid.).

There was competition from school-building that had taken place by the 1840s, which was affecting some mechanics' institutes. At Lockwood, for example, located in a suburb of Huddersfield, by 1847 there had been a decline in membership, due to a 'school being opened up in the neighbourhood'. The Institute's membership was also affected as 'the Huddersfield Institute being near; having a greater variety of classes' attracted members away from the suburb

Table 9.3 Huddersfield and District Cluster of Yorkshire Union of Mechanics' Institutes relating names to numerical locations in Figure 9.3

1 Hipperholm	12 Kirkheaton	23 Milnsbridge	34 Clayton West
2 Liveredge	13 Holywell Green	24 Crossland Moor	35 Skelmanthorpe
3 Heckmondwike	14 *Huddersfield College	25 Slaithwaite	36 Shepley
4 Brighouse	15 Lindley	26 Netherton and Armitage Bridge	37 Meltham and Meltham Mills
5 Sowerby Bridge	16 Lascelles Hall	27 Emley	38 Netherthong
6 Elland	17 Clough Head	28 Dogley Lane	39 Wooldale
7 Ravensthorpe	18 Lockwood	29 Marsden	40 Holmfirth
8 Mirfield	19 Almondbury and Hillhouse	30 Honley	41 Delph
9 Deighton	20 Longwood	31 Kirkburton	42 Dobcross
10 Greetland	21 Linthwaite	32 Farnley Tyas	43 Hinchcliffe Mill
11 Stainland	22 Golcar	33 Shelley	44 Mossley

* Huddersfield College was a private establishment providing several staff who taught at a number of mechanics' institutes in the area.

(ibid.: 61). The Committee at Lockwood did, however, report in 1849 that there had been an increase in membership and volumes in the library, despite the success of the nearby Huddersfield Institute (*Eleventh Annual Report of the Yorkshire Union of Mechanics' Institutes* 1849). The Institute noted that:

> yet when we take into consideration that our Institute is situated within about one mile of Huddersfield, whose flourishing Institution offers advantages which we do not possess, and that ours is composed entirely of working men, we do not wonder that our increase is small, nether do we regret young men going to the Institution which offers them the greatest advantages (ibid.: 68).

Thus, Lockwood, while it accepted that there was a larger institute in the town, and could never compete with the Huddersfield Institute, was content to provide elementary education and accepted that its successful students could then progress to advanced-level courses offered in Huddersfield.

The Institute had seen a fall in membership due to the 'commercial depression' in 1862 (*Twenty-Fourth Annual Report of the Yorkshire Union of Mechanics' Institutes* 1862: 116) and again in 1877 when, despite 'the great commercial depression' it continued to be successful, with a total membership of 305 males and females. This was probably as a result of offering relevant subjects, especially for men, as the most successful and popular subject was inorganic chemistry, which the Committee had decided to offer at advanced level: 'anxious to encourage the study of this useful science, and wishing to keep the young men connected with the Institution, they felt justified in going to the expense of providing the necessary apparatus' (*Thirty-Ninth Annual Report of the Yorlshire Union of Mechanics' Institutes* 1877: 138). Thus, the Lockwood Mechanics' Institute had

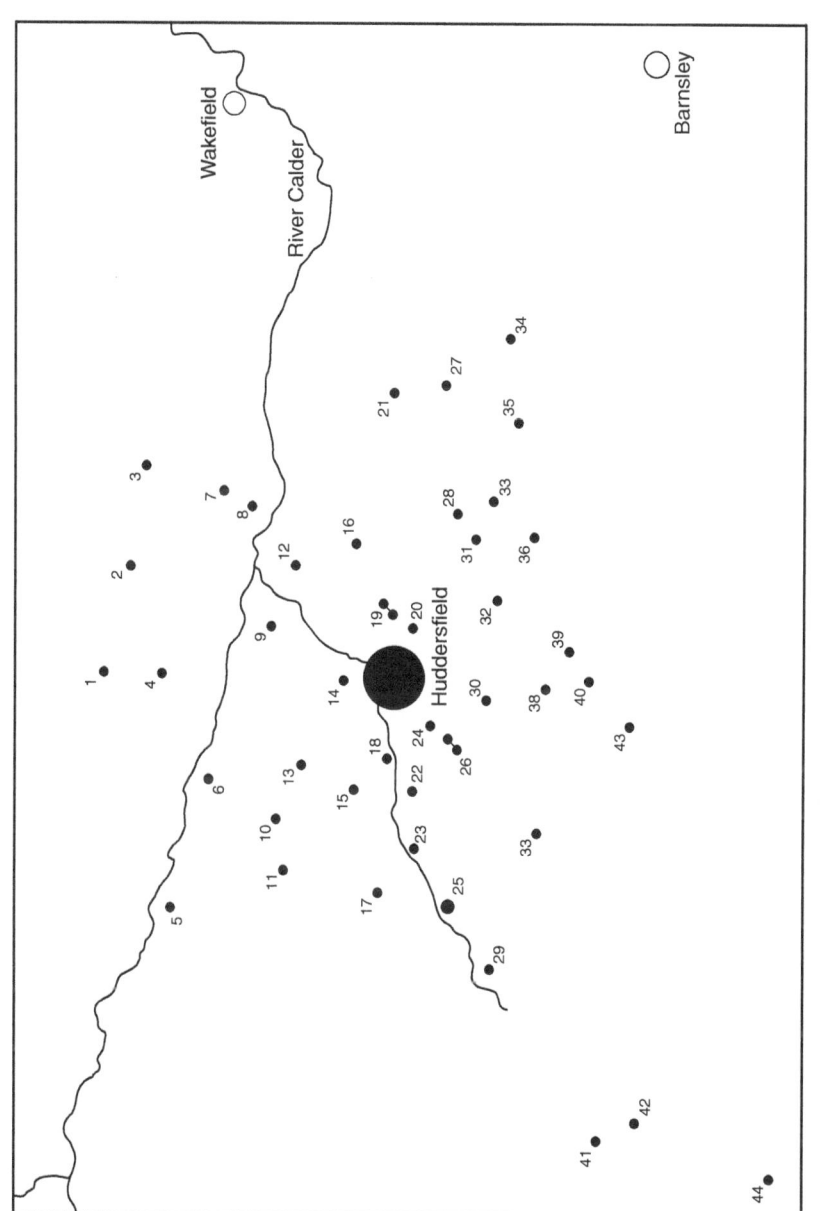

Figure 9.3 Huddersfield and District Cluster of the Yorkshire Union of Mechanics' Institutes.

taken the strategic approach that in order to keep members during economic downturns it was important to offer attractive and relevant subjects that would support their members' employability.

In order to support advanced-level relevant subjects at smaller institutes, committees had to rely on experienced teachers from larger institutes. In 1880, a free-hand drawing class at Lockwood was delivered by Henry Burrows, and a chemistry class by James Allot, both from the Huddersfield Mechanics' Institute. Allot also taught magnetism and electricity, as well as sound and light, at the same institute (*Forty-Second Annual Report of the Yorkshire Union of Mechanics' Institutes* 1880: 113). Being in a suburb of Huddersfield, staff from the Mechanics' Institute in the town who were involved in teaching at the smaller ones would not have had far to travel, making this a convenient arrangement without too much additional time or expense.

It was not only the male mechanics' institute in Huddersfield that attracted some members away from other institutes. The Committee at Almondbury, another institute located in a suburb of Huddersfield, reported in 1865 that, while the Institute continued to be prosperous, it was disappointed that more females did not attend the classes. The Institute was, however, within walking distance from the Female Institution in Huddersfield, which was successful (*Twenty-Seventh Annual Report of the Yorkshire Union of Mechanics' Institutes* 1865).

The exchange of teachers was a common feature of the district. At Lindley, four miles north of Huddersfield, with a population of 4000 inhabitants in 1861 (*Census Returns* 1861), the Committee reported in 1873 that the Institute was in a prosperous state and that 'the educational work has been carried out with vigour and the results quite gratifying' (*Thirty-Fifth Annual Report of the Yorkshire Union of Mechanics's Institutes* 1873: 143). The evening classes were delivered by 'certificated teachers, the success of the students being as a result of their [teachers'] experience'. Many of them taught at other institutes, including at Huddersfield (ibid.: 144). The Lindley Committee, however, noted that students, on completing successfully elementary-level examinations, were then 'continuing their studies in the advanced and practical work at the Huddersfield Mechanics' Institute' (*Fiftieth Annual Report of the Yorkshire Union of Mechanics' Institutes* 1888: 61), as had also been the case at Lockwood for several years. The growth of membership at the Huddersfield Institute had now a sphere of influence outside its immediate environs. In any case, realistically, Lindley and others nearby would not have been able to offer advanced-level courses with such small memberships in the smaller communities. Added to which, the nationally recognised Huddersfield Institute had some of the most advanced science laboratories and curricula in the country. The Institute provided advanced-level courses, particularly in design and science, for students, some of who came from smaller establishments where there would not have been enough students to make these classes viable or the staff expertise to deliver them. There was some competition, but this seems to have been rather isolated in only one or two examples.

As was the case of the institutes in the surrounding villages and suburbs, trade depressions affected the well-established male Huddersfield Mechanics' Institute.

The Committee reported in 1859 that there had been a decrease in numbers due to a trade depression. As there had been a fall in revenue from £426 to £353 there had not been the finances to purchase books either for the classes or the library (*Twenty-First Annual Report of the Yorkshire Union of Mechanics' Institutes* 1859: 89). By 1862, the Committee had decided to introduce fortnightly fees, so members could pay in small regular instalments and not feel committed to paying quarterly, with a much larger outlay (*Twenty-Fourth Annual Report of the Yorkshire Union of Mechanics' Institutes* 1862: 132). However, the town and institute were not affected to the same extent as the smaller communities due to the diversification of industry, including coal and engineering.

In 1863, with an improvement in trade taking place, the Huddersfield Female Mechanics' Institute Committee identified that the fall in membership was due to 'the large demand for female labour' which had been 'the most influential and immediate of the causes', as there was demand to increase production in the local mills following the trade depression (*Twenty-Fifth Annual Report of the Yorkshire Union of Mechanics' Institutes* 1863: 93). In 1879, however, there was a further decline in membership at the Female Institute 'as a result of the continued depression'. This had meant that many of the females still attending had their fees paid by benefactors, among them Mrs Frederic Schwann, the wife of the President of both Huddersfield Institutes (*Report of the Huddersfield Female Education Institute* 1879: 1).

While those institutes in and around Huddersfield were affected by trade cycles, smaller mechanics' institutes some miles outside the town were also affected. Heckmondwike in 1844 had declined in comparison to previous years, and this was attributed to 'the state of the trade' (*Sixth Annual Report of the Yorkshire Union of Mechanics' Institutes* 1844: 25–6). The New Heckmondwike Mechanics' Institute, opened in 1873, was also affected by the 'bad state of trade' in 1876, and had debts of £150. However, much of it was paid off by benefactors, including the Heckmondwike Industrial Co-operative Society which donated £20 (*Thirty-Eighth* Annual Report of the Yorkshire Union of Mechanics' Institutes 1876: 153). The Committee reported again in 1880 that, due to the 'long continued trade depression', membership had fallen and the result was that all classes had to close, with the exceptions of English, writing and art (*Forty-Second Annual Report of the Yorkshire Union of Mechanics' Institutes* 1880: 100).

As was the case in the Dales and Pennines, some of the smaller institutes had also been affected due to rural depopulation. In 1847, for example, Honley Institute was continuing to operate, but the Committee reported that several members had left due to not having the commitment to persevere with their studies, while others had left as a result of 'the want of employment, which has unhappily prevailed to some extent in this neighbourhood, depriving many of the labouring class of the means by which they might obtain the advantages of this Institution' (*Ninth Annual Report of the Yorkshire Union of Mechanics' Institutes* 1847: 44). Rural depopulation was common during this period as families left the rural areas looking for work, and this contributed to the fall in membership of the rural institutes during periods of depression.

The Committee at Holmfirth had reported in 1848 that the majority of its members were of working class and, therefore, due to the 'severe commercial depression', there was a decline in numbers from 162 to 98 (*Tenth Annual Report of the Yorkshire Union of Mechanics' Institutes* 1848: 48). Twelve years later, with improvement in trade, the Committee reported that attendance and membership was again on the decline, as in the case of the Female Institute in Huddersfield: 'owing to the commercial prosperity of our manufacturers, many of that class are debarred from the benefits of such institutions by having to work overtime' (*Twenty-Second Annual Report of the Yorkshire Union of Mechanics' Institutes* 1860: 90). In other words, when times were good employees were expected to stay on and work, indicating that membership was affected negatively in good times as well as bad.

Thus, despite the growth in the number of mechanics' institutes during the period of study, not only did the first generation of institutes (pre-1850) continue to be successful but also new ones that were established within a few miles of Huddersfield. Many continued to operate as institutes and libraries into the early twentieth century, despite being close geographically to Huddersfield and being affected by trade depressions and competition. Indeed, having a particularly well-established and successful institute in the town provided the opportunity for members attending other institutions to benefit from specialist teachers who taught in more than one establishment, and provided an opportunity to progress to higher-level subjects if their own institute was unable to deliver them.

The purpose of selecting such a cluster as the one that developed around Huddersfield was to examine a densely populated area of mechanics' institutes and identify what the impact was on membership patterns. It was assumed that membership numbers in the areas around Huddersfield would be quite small, as the town's institutes were dominant. In fact, those that were located only a few miles from Huddersfield had healthy numbers in comparison to those in larger towns in the other two clusters. Those mechanics' institutes located within only a few miles of Huddersfield are listed in Table 9.4.

Almost all institutes sampled in this cluster increased in size relative to population whether they were close to Huddersfield or a few miles outside of it, which inevitably meant rural areas.

None of the institutes seem to have increased in size in relation to population growth between 1850 and 1860. However, there was an increase in the membership of several institutes in and around Huddersfield, including those at Slaithwaite, Brighouse and Shelley, between 1860 and 1870. Several institutes close to Huddersfield had increases in membership between 1870 and 1880, including those at Holywell Green, Longwood and Lockwood. The growth pattern continued up to 1890 (*Annual Reports of the Yorkshire Union of Mechanics' Institutes* statistical data).

Conclusion

Overall, the research carried out into these particular clusters has provided evidence that several urban, semi-urban, rural and mining communities were

Table 9.4 Selected Mechanics' Institutes for Huddersfield and district, showing ten-year membership

Mechanics' Institute	1850	1860	1870	1880	1890
Almondbury	—	90	—	—	—
Brighouse with Rastrick	142	89	216	209	—
Dogley Lane	82	—	—	—	—
Elland	61	—	—	—	—
Gomersal	108	239	227	86	—
Greetland and West Vale	—	—	—	99	224
Heckmondwike	86	20	—	100	—
Holywell Green	—	—	84	200	110
Holmfirth	172	163	—	—	274
Honley	105	94	30	—	—
Huddersfield (Male)	887	1207	1453	1651	1255
Huddersfield (Female)	127	254	—	—	—
Kirkburton	—	—	—	—	88
Kirkheaton	48	—	—	—	—
Lindley	—	175	119	87	135
Lockwood	51	213	327	296	210
Longwood	—	147	—	29	95
Marsden	147	150	120	—	—
Meltham	154	120	176	—	74
Shelley	—	126	84	105	—
Slaithwaite	—	45	70	80	550

Source: *Annual Reports for the Yorkshire Union of Mechanics' Institute.*

supported by the Yorkshire Union of Mechanics' Institutes, in relation to adult working-class education. While the three clusters had similar features, which were often the same as those across the Union, they also had significant differences, and their adaptability to the needs of their particular communities in relation to their socio-economic environments was substantial. The North East cluster included several general institutes as well as smaller ones, such as those founded by Joseph Pease and family. The more dispersed cluster of the Yorkshire Dales and Pennines included general institutes of the industrialising towns, such as the one at Keighley, as well as smaller ones associated with local upland rural communities. Finally, the densely concentrated cluster of over 40 institutes within only a few miles of the town of Huddersfield, which had the second-largest institute in the Yorkshire Union, has provided insight into how institutes were able to continue to operate in close proximity.

References

Census Returns (1861).

Huddersfield Female Education Institute (1879) *Report.*

Ingle, G. (1997) *Yorkshire Cotton: The Yorkshire Cotton Industry, 1780–1835* (Preston: Carnegie Publishing).

Kirby, M. W. (1984) *Men of Business and Politics: the Rise and Fall of the Quaker Pease Dynasty of North East England, 1700–1943* (London: George Allen and Unwin).

www.legislation.gov.uk (accessed October 2015).

Marsden, W. E. (1977) 'Historical geography and the history of education', *Journal of the History of Education Society* Vol. 6, No. 1, 21–42.

Sanderson, M. (2007) Educational and Economic History: The Good Neighbours, *Journal of the History of Education Society* Vol. 36, Nos. 4–5, 429–445.

Stockdale, C. (1993) *Mechanics' Institutes in Northumberland and Durham 1824–1902*, unpublished PhD thesis (Durham: Durham University).

Yorkshire Union of Mechanics' Institutes (1844) *Sixth Annual Report.*

Yorkshire Union of Mechanics' Institutes (1847) *Ninth Annual Report.*

Yorkshire Union of Mechanics' Institutes (1848) *Tenth Annual Report.*

Yorkshire Union of Mechanics' Institutes (1849) *Eleventh Annual Report.*

Yorkshire Union of Mechanics' Institutes (1851) *Thirteenth Annual Report.*

Yorkshire Union of Mechanics' Institutes (1852) *Fourteenth Annual Report.*

Yorkshire Union of Mechanics' Institutes (1859) *Twenty-First Annual Report.*

Yorkshire Union of Mechanics' Institutes (1860) *Twenty-Second Annual Report.*

Yorkshire Union of Mechanics' Institutes (1862) *Twenty-Fourth Annual Report.*

Yorkshire Union of Mechanics' Institutes (1863) *Twenty-Fifth Annual Report.*

Yorkshire Union of Mechanics' Institutes (1865) *Twenty-Seventh Annual Report.*

Yorkshire Union of Mechanics' Institutes (1873) *Thirty-Fifth Annual Report.*

Yorkshire Union of Mechanics' Institutes (1874) *Thirty-Sixth Annual Report.*

Yorkshire Union of Mechanics' Institutes (1876) *Thirty-Eighth Annual Report.*

Yorkshire Union of Mechanics' Institutes (1877) *Thirty-Ninth Annual Report.*

Yorkshire Union of Mechanics' Institutes (1879) *Forty-First Annual Report.*

Yorkshire Union of Mechanics' Institutes (1880) *Forty-Second Annual Report.*

Yorkshire Union of Mechanics' Institutes (1888) *Fiftieth Annual Report.*

10 Expansion of the Mechanics' Institute Movement beyond Britain

The Mechanics' Institutes were not only to be found in Britain but were also established overseas. By 1850, Hudson (1851) identified that several had also been founded in various parts of the world including Germany, where at Hamburg a mechanics' institute had been opened in 1848 'for the instruction of labourers, by means of evening classes'. However, there was such a well-established education system for all that mechanics' institutes were not found in such large numbers as in Britain. India had several, amongst them the Calcutta Mechanics' Institute, which offered lectures in physical science, manufacturing, commerce and agriculture, and Bombay Mechanics' Institute offered courses in science, manufacturing, printing and lithography, pottery and metallurgy. Mechanics' institutes were also founded in Canada, holding exhibitions 'of a practical nature'. They were established in Montreal, Quebec, Hamilton and Toronto, and by 1850 Niagara and Amherstburgh also had mechanics' institutes (Hudson: 219). New Zealand had a mechanics' institute at Port Nicholson, founded in 1842, with a reading room and library, which Whitelock (1974: 87) believes became 'the forerunner of scores of mechanics' and literary institutes in that colony'. In 1850 the Wellington Athenaeum and Mechanics' Institute, which had a museum and library, was opened in the same country (Hudson: 221).

Bode (1956) makes the specific point that the British Mechanics' Institute Movement influenced observers overseas including those in France. Charles Dupin, having visited the Glasgow and London Institutes several times, undertook the challenge of convincing the French government of the importance of offering working-class adult education through providing state financial support. Institutes were established in La Rochelle, Nevers, Lyon, Mertz and Versailles. Dupin publically thanked the government for their support of the Movement which, by the end of 1826, had reached 98 towns, which 'were endeavouring to rival each other in their zeal for imparting the new instruction to the working classes' (Bode 1956: 7) With so many developments taking place abroad, this chapter will concentrate on the origins of American and Australian Mechanics' Institute Movements.

The Movement in America

It has been assumed that the inspiration of establishing mechanics' institutes in America seems to have been based on the British Mechanics' Institute Movement.

They were certainly familiar with the influence of Birkbeck and Brougham and their contributions in supporting education of the working class. The first American societies for the diffusion of useful knowledge were similar in organisation and use to the first British example, founded by Brougham in 1826 (Kett 1994: 112).

It is also known that American newspapers carried European news stories about politics and developments including the Mechanics' Institute Movement. The speeches of Brougham, for example, would have been put into print through magazines and periodicals In Britain; it was the *Mechanics' Magazine*, first published in 1823, which provided regular features and discussed developments. In America, the equivalent was the *American Mechanics' Magazine*, which was first published in 1825. Most of the content in the *Magazine* seems to have been based on British– and London–related articles. The following year, the periodical was renamed and rebranded the *Franklin Journal and American Mechanics' Magazine*. However, this journal seems to have become too scientific for a readership likely to attend the local mechanics' institute or lyceum. Instead, the *American Journal of Education* published in its third issue a substantial article on British mechanics' institutes, suggesting it was encouraging America to establish its own Mechanics' Institute Movement (Bode 1956: 9–10). Sinclair (1974) makes reference to the Franklin Institute of the State of Pennsylvania for the Promotion of the Mechanic Art (hereafter Franklin Institute) stating that it is possible to trace the society's beginnings to the Mechanics' Institute Movement from both Britain and within America, suggesting that the aims behind the British movement of providing knowledge 'should be open to all that seek it; learning should be restricted neither by closed societies nor by exclusive universities' (Sinclair 1974: 5). To many in America, Sinclair (1974) believes that

> Birkbeck was the central figure in the drama. And Americans, who always felt comfortable in the assurances of old-world precedent, Birkbeck, Glasgow, and London, were the proofs they urged in support of their own ambitions. They might just reasonably have looked for examples amongst themselves. The mechanics' institute began simultaneously in Britain and the United States. The London Institute and the Franklin Institute were established practically at the same time (ibid.: 7).

Thus, there were developments in both countries that suggest the movement expanded simultaneously. Like Birkbeck in Britain, John Griscom is seen as the founder of the movement in America. The New York Mechanic and Scientific Institute was established in 1822, and one of its co-founders was Griscom. Interestingly, both men were from Quaker families (Kett 1994: 112).

Tylcote (1957) states that in July 1823 the Liverpool Mechanics' Institute and Apprentices' Library was opened, which was influenced by the New York Apprentices Library that had been opened in 1820 (see Figure 10.1). A 'painted silken banner, sent by the apprentices of New York to those of Liverpool, was unfurled, and it had both the eagle and lion and inscribed with the message

Figure 10.1 The successor to the Liverpool Mechanics' Institute and Apprentices'
        Library, which was opened in 1825, was the Mechanics' Institute
        and School of Art, built 1835, and is now the Liverpool Institute of
        Performing Arts.

Source: Author's collection.

New York sends her good wishes to Liverpool' (*Mechanics' Magazine* 15 November
1823). A gift of 30 volumes of English writers, printed in the Union, was sent
to Liverpool from New York, Philadelphia, Boston, Baltimore and Connecticut
for the Liverpool Apprentices' Library. This American influence is of course not
surprising, with Liverpool's transatlantic connections with New York (Tylecote
1957: 55–6).

Writing about the founding of mechanics' institutes in America, Stevens (1990: 528) described them promoting practical and scientific knowledge and supporting an 'upward mobility of that broad class of workers called mechanics', and were the 'great democratic leveller ... for social mobility'. For example, it was noted by a contemporary in 1836 that mechanics in Philadelphia were now able to read, reflect and investigate. As in Britain, by the mid-nineteenth century, science was popularised in American institutes through evening lectures and classes that encouraged not only the skilled mechanics but also apprentices and working men (ibid.: 530). In 1834, the State Convention of Mechanics' Institutes held in New York listed the following trades under the general term of mechanics. They included tailor, cooper, silver plater, cordwainer, ironfounder, chair and cabinet worker, hatter, saddler, blacksmith, machinist, watchmaker, tanner, tin plate worker, stone-cutter, carriage-maker, brassfounder, brushmaker, printer, publisher, pianoforte maker, locksmith, marble-cutter and carpenter (ibid.: 528).

In 1832, the American Mechanics' Institute in New York was established. It was founded by a group of manufacturers and several professionals. Like its British counterpart, the Institute did not permit political activities and instead concentrated solely on providing education for its members. An unnamed committee member at the time stated that without education in science, ethics and economics, the 'laboring classes of the community will be doomed to an intellectual and political slavery by the better educated classes' (Wilentz 1984: 272). By 1836, the Institute had over 500 members, with an extensive lecture series. It had its own journal, with the use of rooms in the basement of the Civic Hall. The Institute was seen as being run on similar lines to the Birkbeckian London Mechanics' Institute as both were attended by manufacturers and employers (ibid.: 227). Its membership was over 1,300 with a small group of professionals, but the majority were journeymen covering a variety of trades including 'stone cutters, printers, machinist, tailors, blacksmiths, painters, carpenters, shoes makers, hatters and cabinet makers' (Rice 2000: 279). Like those in other parts of the world, the New York Institute stated in its constitution that 'nothing of a religious, irreligious, or political tendency shall be admissible on any account, at any meeting of the Institute' (ibid.: 283).

There were also the New York Apprentices' Library and Mechanics' School, both of which were founded in 1820. The school was mainly attended by children of deceased members and those who could afford the fees (Wilentz: 88). In 1826, the Mechanics' School and Apprentices' Library were opened to girls and, by 1830, with over 200 pupils, the staff had to adopt the monitorial system similar to that commonly used in Britain (ibid.: 149).

During the late 1820s and early 1830s both Britain and America had a common supporter in Timothy Claxton. He was a self-educated London artisan who promoted mechanics' institutes and similar institutions in both countries. He studied mathematics and drawing in his own time and had built several mechanical instruments. Claxton had tried to establish a philosophical society for mechanics but it failed. He also applied for membership of an established society but was turned down, as he was 'only a mechanic'. He moved to America during the 1820s and in 1826 established the Boston Mechanics' Institution. By the 1830s he was co-founder of the Boston Mechanics' Lyceum, established especially for artisans (Kett 1994: 113).

As previously mentioned, mechanics' institutes in Britain were often supported by professionals, industrialists, merchants and others with close connections with their respective communities and employees. This was also true of America. For example, Samuel Vaughan Merrick, who realised, having inherited a bankrupt mill, that he had little technical knowledge, and called upon relatives, one who had studied under the chemist Joseph Priestly in Britain, to support the foundation of the Franklin Institute in Philadelphia for the purpose of developing his and others' workforce to support industrial needs (ibid.: 114). The Franklin Institute was founded in 1824 and Sinclair (1974: 3) states that it was modelled and organised on the 'mechanics' institutes of Great Britain'.

The Working Men's Institute of New Harmony in Indiana was founded in 1838 by William Maclure, who was born in Ayr, Scotland, in 1763. He travelled throughout Europe and then visited America in his capacity as a British merchant. In 1800 Maclure became an American citizen, where he had become a Fellow of the American Philosophical Society. While on a visit to England, he had visited Robert Owen's cotton mills and New Lanark model village, and believed that something similar for the education of adult learners available to those in New Lanark should be established back in America (Doug 1991: 404). Maclure would have been aware of the British mechanics' institutes and their libraries (ibid.: 406).

Owen decided to visit America and introduce his philanthropy and New Lanark Model at Harmony, situated on the River Wabash in Indiana. Indeed, Owen bought the village from a German religious sect that wanted to return to Pennsylvania, in 1825. Maclure had some reservations about Owen's plan but agreed to be responsible for education affairs at what Owen had renamed 'New Harmony'. In 1827, Owen left, having won a lawsuit over financial arguments with Maclure. Maclure wanted to set up an institute similar to that in Britain and in 1838 he founded one in New Harmony, which he hoped would encourage a movement across the whole of America (ibid.) The New Harmony Working Men's Institute for Mutual Instruction was opened on 2 April 1838. The aim of the Institute was to bring scientific and useful knowledge within the reach of labourers through reading, lecture and scientific demonstration (ibid.: 407).

Within a year of its opening, the Working Men's Institute had a reputation as a successful public library that was open to members and their families, including children (ibid.: 409). Doug (ibid.: 412) states that the Institute, having a successful library, was a model that supported the successful public libraries in America, for example, at Boston in 1850 and the American Library Association in 1876.

Goldsmith (1955: 2) states that technical education in California was first introduced at the San Francisco Mechanics' Institute in 1855. With the gold rush over and trade declining, mechanics in the town were concerned that industrial prospects were poor. The solution was to make updated technical knowledge available to them and others across all trades and industries. The original committee who discussed the formation of an Institute were made up of a machinist, foundry builder, stonemason, foundryman, carpenter, merchant and sawmill owner. The constitution stated that a library and reading room, scientific apparatus and works of art for scientific purposes would be provided for members.

The Committee rented rooms on the fourth floor of the Wells Fargo Express Building and, while its first concern was to establish a library, it was not long before classes were being delivered for mechanics, developing projects promoting local industries and planning to fund and erect a building of their own. Free public lectures were delivered on technical developments (ibid.: 5–6).

In 1857, the Institute Committee decided to hold a Mechanics' and Manufacturers' Fair[1] in order to promote industry and raise revenue. Land was donated and a pavilion built at the cost $7000 to house the 941 exhibits. The fair made a profit, some of which was donated to two orphanages and the rest of which was reinvested to hold further fairs over the coming years (ibid.: 9).

Following the end of the Civil War, which hindered mechanics' institutes generally, and with the opportunities to expand industrial growth, the San Francisco Institute responded by offering classes in the new technical developments that were taking place. At the same time, gold mines in California were expanding and there were more patents being lodged with the State than in any other in America. The majority were for mining machinery and ore extraction, agriculture and milling. The Institute's involvement in providing classes in mechanics and mining contributed to this success. Indeed, the legislation in support of establishing the University of California in 1868 made specific reference to the Mechanics' Institute, recognising its contribution to technical education. It was at this time that the Institute moved into its last, permanent building. Annual fairs continued to be held in more elaborate pavilions, the sixth one built in 1886 being particularly impressive (ibid.: 13). It also lobbied for a railroad, which, following its opening, made substantial contributions to the city and its environs, both socially and economically (ibid.: 15).

Throughout the 1870s and 1880s, the Institute was offering mechanical drawing, applied mathematics, wood-carving, ironworking, free lectures and annual fairs. It was also offering what Goldsmith (ibid.: 16) refers to as general education and what Britain would refer to as elementary education. The University staff delivered weekly lectures at the Institute. With the income from the fairs, often exhibiting works from China, Japan, Australia, as well as locally made San Francisco mining equipment, the Institute was able to expand its book stock. The final fair was held in 1899, by which time they had served their purpose.

At the time of the San Francisco earthquake in 1906, the Institute had 200,000 books. Following the tragedy, the building was destroyed, with all its books and archives, including British patent reports going back to King James 1 as well as technical and scientific papers and works of art. At the time of the fire, the librarian was already contacting book dealers and libraries in the eastern states by telegram, asking if it was possible to send books, especially on architecture and engineering, for restocking in what would be a temporary building, some five months after the earthquake. In 1910, a new nine-storey purpose-built institute was opened and by 1912 it had over 50,000 books in the library (ibid.: 24).

Scientific and technical courses continued to be important and were reflected in the new energy industries of oil and hydroelectric power as well as automobile manufacture. During the two World Wars, the Institute continued to support technical research and new business developments required for the twentieth century. By the 1950s, the Institute had a library stock of over 155,000 books and documents, and over 5,000 members (ibid.: 28).

Erastus Bigelow had wanted to go into higher education after attending a local school in West Boylston, Massachusetts, and an academy in Leicester. His father, however, refused to support this and expected him to work as a mill-hand at the small family cotton mill. He seems to have hated the work, and left home and went to Boston to make money to support his higher education. He taught himself stenograph (shorthand) and both taught and wrote a book on it, which was popular in Boston. Bigelow also designed and developed a power loom for weaving coach lace used for horse-drawn carriages. With his older brother Horatio, and other colleagues, he developed a factory village in Lancaster. In 1840, the Bigelow Mechanics' Institute was founded and funded by Horatio and others who believed in supporting those who were committed to science in support of invention and industrial developments (Kett 1994: 114–17).

Like, many institutes in Britain, the Bigelow Institute tended to concentrate on advanced science courses, which many found hard to understand. 'Words like oxide and alkali meant nothing to them (artisans), and many attended lectures on science only to witness the dazzling lights and deafening explosions that accompanied demonstrations of chemistry' (ibid.: 117). However, this and others, notably the Franklin Institute and Ohio Mechanics' Institute, did not prevent expansion of the movement, and many patrons saw them as keeping 'apprentices and journeymen out of taverns and others habitats of vice' (ibid.: 117–18). This, too, was seen as an aim of the British mechanics' institutes, particularly those supported by the Temperance Society or other tee-total groups and individuals. Thus both countries had concerns that those for which their respective movements were founded were not attending, and instead those of higher social rank and the professions had taken them over for their own needs.

As in Britain, institutes in America began offering science that supported working men. As the *Mechanics' Magazine and Register of Improvements* put it, 'there is hardly any trade or occupation in which useful lessons may not be learnt by studying one science or another' (ibid.: 118).

While in Britain during the 1860s the Mechanics' Institute Movement was reforming in support of working-class education, in America similar institutes 'gradually sank into insignificance' (ibid.: 119). This divergence between the two countries was not due to differences in their aims and ideologies in support of working-class education and training. Instead, it seems to have come about through political ideology in America, where working men's parties were established to support their members' employment rights and had little interest in diffusion of knowledge for all.

Secondly, during the second half of the nineteenth century, British mechanics' institutes continually adapted to supporting vocational education and training, including elementary education for men, women and children, as well as establishing technical schools and schools of design, all of which were run as voluntary organisations and charging fees. In contrast, American Institutes seem to have reverted back to scientific centres of learning for those who had attended schools. In any case, the majority of people in America had access to free schools. However, several institutes did open their doors to children, among them The General Society of Mechanics and Tradesmen in New York, the Ohio Mechanics' Institute (which also had a School of Design) and the Franklin Institute, which launched its own separate high school (ibid.: 121).

Added to this was the fact that, with the increase in popular science, higher fees could be charged than those afforded by artisans and other similar groups. Thus, it was only in the 1890s that a movement for mass vocational education started to develop in America, some 40 years after doing so in Britain following the Great Exhibition.

Where there was a distinct difference between America and Britain was that some American institutes reinvented themselves as lyceums, offering lectures in science and attracting members of the public who had more than a passing interest in popular science, rather than scientific knowledge in support of their trade in industry (ibid.: 125). They seem to have been more popular in America than Britain where, in the case of the latter, they were generally associated with places of entertainment. However, Bode (1956: 7) believes that the lyceums were based in the British movement. What made them different was that America offered practical rather than scientific education through lectures. Scientific knowledge was more associated with the Mechanics' Institutes Movement in both countries. The term 'lyceum' was often chosen to depict them as neutral or universal and used to indicate inclusivity rather than a particular social class of member. 'Whereas organizational names like Mechanics' Institute, Mercantile Library Association or Young Men's Association imply that the societies target mechanics, mercantilists, or young men, a name like Davenport Lyceum … implies a restriction on participation based only on geographic area' (Ray 2005: 4). Like mechanics' institutes in America, lyceums were seen as supporting mental improvement and preventing young men moving into towns to work in the mills and factories from 'drinking, gambling, fighting, hiring prostitutes' (ibid.: 17).

However, Sinclair (1974: 13) points out that mechanics' institutes, lyceums, apprentices' libraries and other similar institutions had common aims and often, where more than one was situated in the community, 'overlapping membership', and therefore could easily merge when necessary. In any case, he identifies that mechanics' institutes were not established in the West and the South until the 1840s and 1850s.

Another similarity with Britain was that the respective movements were influenced by individuals. In Britain, it was Birkbeck and Brougham, in France, Duplin and in America, Josiah Holbrook (Bode 1956: 8). By 1828 the Lyceum Movement, believed to have been the work of Holbrook, was spreading across

New England, with around 60 such institutions being established. The number had increased to 5000 by 1839, with the movement spreading from New England to Missouri and Florida. In Massachusetts alone there were 137 lyceums. Other states had their fair share of lyceums, too (Ray 2005: 21). Lyceums resembled mechanics' institutes, and the first one was established in 1826 (Whitelock 1974: 87). To begin with, many lyceums were in public places such as church halls, schools or town halls, but later many operated from purpose-built halls. Such buildings seem to have had a lecture hall, classrooms and a cabinet room. The latter was a repository for mineral collections and library books (Ray 2005: 22–3). While the lyceum seems to have been at the heart of village or town community life through offering both general and specialist individual lectures, by the 1880s they were in decline. Those that survived tended to have done so through offering their hall for social gatherings and entertainment (Ray 2005).

By the mid-nineteenth century, some towns seem to have had a lyceum while others had a mechanics' institute, both offering practical education to their membership in their own ways, the former often but not exclusively theoretical and the latter practical. In 1843, the Chicago Mechanics' Institute seems to have encouraged travelling lecturers and, in return for giving classes, they were given accommodation for free. Members were admitted for free; others paid an entrance fee to cover the costs (Bode 1956: 97).

In 1826 James Smithson, an English scientist, left $500,000 in his will to the United States government to establish and fund an institution 'for the increase and diffusion of knowledge' in Washington, D. C. In 1846, the Smithsonian Institution was opened, and the building housed a museum with geological and mineralogical specimens, a chemistry laboratory, art gallery, lecture rooms and a library. Over a period of time, the Smithsonian libraries expanded into 22 branches, located in various Smithsonian institutes in Washington D. C., New York City, Edgewater, Maryland and the Republic of Panama.

Unlike libraries in Britain, which were only slowly becoming public libraries sometime after the passing of the Public Libraries Act in 1850, in America it was common for funds to be left in wills to support free public libraries rather than this being done through the government. However, subscription libraries, factory libraries and mechanics' institute libraries supported access to books for all sections of society in both America and Britain, with the increasing demand for books and the prohibitive cost for many of purchasing their own (Wiegand and Davis 1994: 329–31).

In the case of both countries, libraries were introduced to supplement lectures and classes offered at the institutes. Similarly, both countries increased their library stocks to include recreation and fictional works so as to broaden the range of interest and increase membership. By the later nineteenth century, as in Britain, America saw the decline of such libraries, which were superseded by government-funded libraries on both sides of the Atlantic (ibid.: 331).

As previously mentioned in Chapter 4, British mechanics' institutes often gained additional income and raised public awareness from holding exhibitions displaying art, scientific experiments, machines and other exhibits. These were established before, and continued to be annual events after, the 1851 Great Exhibition held in London. In America similar events were held at the larger institutes annually, although they were often referred to as 'mechanics' and manufacturers' fairs', as in the case of the San Francisco Mechanics' Institute (Bode 1956: 181).

As in Britain, self-improvement was based on useful knowledge, which was offered by the mechanics' institutes. The curriculum also seems to have been similar, with reading, writing and arithmetic being the basis on which elementary education was offered. Higher-level work involved algebra, geometry and trigonometry, navigation, surveying, geography and astronomy. Lectures, often given by travelling academics, were very common (Sinclair 1974: 9).

A further similarity to Britain was that those lyceums or mechanics' institutes that declined or closed often donated their libraries to the town through the public libraries movement. For example, 10 public libraries were founded in Massachusetts between 1851 and 1854 alone. As various institutions fell into decline they often donated their books to their local public free library (Bode 1956: 245).

Australia

The first mechanics' institute founded in Australia was at Hobart in 1827, established by British settlers who must have seen the importance of establishing adult education. There is much evidence that the ties here with the British Mechanics' Institute Movement were even closer than those that Britain had with America, and that its influence was stronger (Whitelock 1974: 95).

At Hobart, master tradesmen formed a committee for the purpose of setting up a mechanics' institute. Lectures on steam engines, and mechanical and engineering science, were given by James Ross from Aberdeen University, and the Institute was known as the 'Birkbeck of Tasmania' (Nadel 1957: 131).

In 1831, the *Stirling Castle* charter ship arrived at Port Jackson, Sydney. Her passengers included 52 mechanics and their families who had sailed from Scotland some four months earlier in support of developing the area around Botany Bay on the outskirts of Sydney. Among the passengers was the Rev. Henry Carmichael, who would be the first pioneer in Australia of adult education. While at sea, he had reputedly turned the *Stirling Castle* into a floating mechanics' institute, providing passengers with the opportunity to be taught arithmetic and geometry, as well as political economy based on the work of fellow Scot, Adam Smith. Indeed, several of the mechanics had been members of the Edinburgh [Mechanics'] School of Arts (Whitelock 1974: 85–6).

The close connections of the mechanics with the Edinburgh Mechanics' School of Arts meant that the institute they founded in Sydney in 1833 was organised on similar lines. Those present at the inaugural meeting, chaired by a Justice of the Peace, included a saddler from Ireland, a boot maker from England and a builder from Scotland (ibid.: 96), They had at the meeting the 'English book of rules' for mechanics' institutes.

The object of the Sydney Mechanics' Institute, or Sydney Mechanics' School of Arts, was 'the diffusion of scientific and other useful knowledge as extensively as possible throughout the colony of New South Wales'. This involved planning for a library, a reading room and apparatus for lectures 'on the principles of physical and mechanical philosophy'. The Committee was also to have been made up of a majority of operatives (Nadel 1957: 118). By the 1870s the Sydney Mechanics' Institute was supported by a wide range of members, not all working class. It included a Technical and Working Men's College, which increased its popularity through offering relevant practical learning that brought new members into contact with the Institute (Dunn 2012). In 1883, the New South Wales government took over the College and re-established it as the Sydney Technical College, but still operated from the Sydney Institute building until 1891 when it moved to purpose-built premises. In 1988, the institution was renamed the University of Technology, and after further developments the buildings became part of the campus of Sydney Institute of Technical and Further Education. During the 1970s the Sydney Mechanics' Institute lost membership as a result of many years of government funding for free public libraries. It sold the original building and moved into new premises across the road in 2000 (ibid.).

Why were mechanics' institutes founded in Australia?

The Mechanics' Institute Movement in Britain was very much a response to the need for education and training for the industrial classes in a period when there was no state education for all. The American movement also came about for the need to support the working population with institutes, lyceums and public libraries. However, Australia in the 1830s had little or no industrial development on the scale of that in Britain and therefore not the same need for technical education. Those colonising the country were 'bush workers, farmers, convicts, or ticket-of-leave men, generally living in isolation in small communities or in tiny outposts of settlement' (Whitelock 1974: 97).

Thus, the main aims of the mechanics' institutes in Australia were very similar to those in Britain. As Carmichael observed, they were to be established:

> To teach the mechanics the theory which underlay the practical occupations in which they were engaged, to instruct them in science rather than in art ... further, the colonial institutes were to supply the general educational deficiencies of those who had emigrated from the mother country with the intention of residing permanently in the colony and who might thus have missed opportunities for education at home (Ryan 1974: 12).

The town of Sydney had only a population of 20,000 inhabitants compared with Manchester, which had 200,000. Yet the Sydney Mechanics' School of Arts, with 600 in 1839, was a third of the Manchester Mechanics' Institute membership (Whitelock 1974: 98).

Following the opening of the Sydney Mechanics' Institute, others were founded in other parts of the country such as Newcastle (1835), Maitland (1838), Adelaide (1838), Melbourne (1839), Launceston (1843), Brisbane (1849) and Perth (1851). From these larger settlements, the movement spread out into the bush towns in the respective areas (ibid.: 104–5). This was very similar to what happened in Britain, with the spread of the movement into the rural and semi-rural areas, particularly but not exclusively in the North of England.

Similarly, the Mechanics' Institute Movement in Australia went into decline by the mid-nineteenth century through subjects being offered that were not relevant to the masses, and membership decreases during periods of economic decline. However, with the expansion in population came the need for more advanced manufacturing and infrastructure. Mining developed and required its workforce to have the skills and expertise associated with the industry. Schools of Mines, based on mechanics' institutes, were opened in Ballarat (1871) and Bendigo (1873) in the State of Victoria. There was therefore an urgent need for technical knowledge to support industrial progress. In 1869, a technical education commission had been set up in Melbourne following the success of the Great Exhibition in London in 1851 and the Paris Exhibition of 1867 (ibid.: 111–12). The Australian gold rush was the equivalent to Britain's industrial expansion, bringing with it population growth, new communities and the need for education and training. From 1862 the government financially supported Victoria institutes as free public libraries and, following the passing of the Education Act in 1872, many institutes were used as schools (ibid.: 125).

While the Mechanics' Institute Movement was vibrant in the States of Victoria and New South Wales, almost the opposite was true of South Australia. While Adelaide had an institute, few others existed. This may have been due to the Nonconformist and Lutheran Churches' influences of providing required secular knowledge as well as religious guidance. However, South Australia did contribute to adult education through the establishment of 154 institutes' 'solid halls' libraries. With the passing of the South Australian Public Library Act of 1884, 'serious' books were donated to the new libraries, and the others continued as subscription libraries (ibid.: 126).

Generally, the planners of the colony of South Australia were both wealthy and radical. They identified that education for the masses was a good thing, and several of them had had connections with mechanics' institutes in Britain (Talbot 1992: 6). By the 1830s, institutes 'were an accepted and expanding element of British cultural life'. During the 1840s and 1850s, as in Britain, the movement expanded across the colony following the opening of the Adelaide Mechanics' Institute in 1838. These included the Hindmarsh and Bowden Mechanics' Institute (1847), Gawler (1848), North Adelaide (1851), Port Adelaide (1851), Burra Burra (1857), Glen Osmond (1854), Sturt (1856) and Willumga Mechanics' Institutes (1854). Several of these, and other institutes, had their own buildings before they were opened, such as Burra Burra, a settlement developed around the copper mining industry, and Glen Osmond, founded by the Unitarian Arthur Hardy who had connections with the movement in London. Willumga, on the other hand, never had its own premises until the 1950s (Talbot 1992).

As was true in other states, while lectures were offered on average two a month, the libraries of these and other institutes became the central focus of their work. Where they had only a small collection of works, travelling box libraries were set up, providing institutes with the loan of books so that their membership thought it worthwhile to continuing paying fees. The South Australian Institute, which came about as a result of the merger between the South Australian Subscription Library and the Adelaide Mechanics' Institute, supported the smaller State institutes with loaning books in this way (ibid.: 27). The book box scheme was the first in Australia, followed not long after by similar arrangements in New South Wales and Victoria. The House of Commons Select Committee on Public Libraries referred to itinerating and travelling libraries, a British initiative which was taken up by those settling in Australia (ibid.: 55).

The Mechanics' Institute Movement dominated Australian adult education for most of the nineteenth century. Its success was in 67 institutes responding to the needs of adult education, resulting in further expansion. By 1870, in Victoria alone, there were over 100 institutes, Athenaeums, public halls and libraries. By the end of the century, there were a further 300 established in the State alone (Whitelock 1974: 124–5). The small ones were known as 'chapel cheapies', similar to the one at Arthurs Creek that was built in 1887 (see Figure 10.2), for their basic design, while the more spectacular institutes, such the one at Ballarat, were referred to as the 'Goldrush Glorious' (*South Bourke Standard*).

There were several factors that explain why the Movement was so successful in Australia. There was the respectability of what was then an emerging British

Figure 10.2 Arthurs Creek is located in a rural community some 60 km north-east of Melbourne. Land for the Institute was donated by a local farmer.

Source: Author's collection.

institution that had had some success in highlighting the importance of practical knowledge in supporting industrialisation. Whitelock (1974: 99) believes that the 'colonial culturists' shared this enthusiasm that they were first made aware of in their homeland. With the beginnings of industrial activities, training schemes were required in support of the new communities, such as in agriculture, brick-laying, carpentry and blacksmithing, thus supplying vital skills. Melbourne, for example, was 'in the heat of an enormous gold rush' by 1854, at the time of its institute development (Stam 2001: 8530). They were also a form of moral reformation, through providing adult education in the community and mental improvement as a distraction from immoral vices.

In the State of Victoria, for example, by 1839, mechanics' institutes had quickly become part of the new European settlements. Apart from providing adult education, many formed the basis and organisation associated with the later developing Schools of Arts, Schools of Design, Schools of Mines and present-day technical colleges (see Figure 10.3).

Such institutes were the first and longest-surviving libraries and reading rooms in the State. Institutes, being established by the local community for the community, were centres for debates on both local and national issues including politics and votes for women. Unlike in Britain, the membership seemed more relaxed and able to discuss important developments of the day, perhaps because they evolved and matured more quickly in the new country where there was little constraint or central control. In many rural areas the institute was the civic centre, its building providing the community with a bank, primary school, art gallery, theatre and even a town hall, until these services had their own buildings (Baragwanath 2000: 70).

The Mechanics' Institute Movement in Australia dominated nineteenth-century adult education. The Public Instruction Act of 1880 resulted in the setting up of evening public schools 'to instruct persons who may not have received the advantages of primary education' (Whitelock 1974: 127). However, as Whitelock states, 'it was the ubiquitous, adaptable, numerous, and popular institutes that ruled the adult education roost. By 1900 there were well over a thousand mechanics' institutes 'speckled across the continent'.

While there were several similarities with Britain, the success of the Australian Mechanics' Institutes Movement was supported by government funds, unlike Britain where all income came from membership, exhibitions and other sources of non-government revenue. The other significant difference is that today many mechanics' institutes are still serving their communities, albeit for social rather than educational purposes. Free public libraries had been established earlier than in Britain and institutes had their own subscription library, many of which continue today. Those in larger towns often became colleges and universities.

Mechanics' Institutes and the Library Movement in Australia

As in Britain, there were commercial or subscription libraries established in Australia by the 1820s, such as the Sydney Reading Room, founded in 1820

(Ryan 1974: 8). This was some seven years before the opening of the first mechanics' institute in Australia at Hobart, Tasmania. With the opening of the Sydney Mechanics' School of Arts, a further library was established in the town, which, by 1860, had 8000 books supporting 1194 members (ibid.: 11). With the passing of the 1867 Municipalities Act in New South Wales, libraries were established separately from those connected to mechanics' institutes, providing they were open to the public, and grants of £100 were available where it was proved that a minimum of 300 people were using it. To receive £200 from the authorities, over 1000 people in the locality had to be using the library. These libraries were seen as superior to those of the mechanics' institutes – not surprisingly, since they had public funding to support their success (ibid.: 13).

The Free Library Movement spread from New South Wales to Victoria and South Australia. The only State that did not offer public libraries in the later nineteenth century was Queensland. Meanwhile, mechanics' institutes were supported through travelling libraries, particularly those located outside the main towns, supported by public library schemes (ibid.: 33). As in Britain, books were delivered to the institutes in boxes that could be converted into bookshelves. They were circulated amongst other institutes over the following months, to be replaced by more books.

As in the case of Britain, the Australian mechanics' institutes were providers of library services. Some are still subscription libraries and several have held on to their nineteenth-century materials. Most stock, however, was handed over to the public libraries or was lost when they fell in disuse after World War Two. However, some continue to lend books. These include the Melbourne, Maldon, Chiltern and Stanley Athenaeums as well as the mechanics' institutes at Footscray, Prahran, Briagolong and Ballarat. In the case of the latter, it has both historical books and archives relating to the Institute, as well as others donated by former institutes (Clancy 2000).

Impact of the Australian Mechanics' Institute Movement on the industrial class

Nadel (1957) believes that the social values of education supported the prosperity of working men in a new country such as Australia. Social mobility in a new country meant that education was vital in supporting the working population in having any opportunities to get on and be successful. Publications of the time, such as *The Moral and Intellectual Culture of the People, Essential to Secure Advantages of High Wages and Political Privileges* (1853) and *The Intellectual Opportunities of the Working-Man in Victoria*, the latter published in the *Melbourne Journal* in 1857, supported the aims of the Mechanics' Institute Movement. Indeed, Carmichael gave a lecture at the Maitland Institute in 1857, passionately emphasising how crucial it was for working men to attend this and similar institutes, giving the example of how tradesmen rose to the position of small capitalists as a result of their education, 'leaving room for others to take their places and rise themselves in turn'. The Sydney School of Arts Committee stated that the ability of the colonist to 'force himself upwards, even to the top of the social scale' through education, was similar to what other mechanics' institutes were also proclaiming (Nadel 1957: 173–5).

The movement was further helped in supporting education of the working population through the introduction of the Eight-Hour Working Day Act in 1856. Following its introduction, so allowing time for study, there was an increase in the number of mechanics' institutes between 1856 and 1857. The Secretary of the Sydney School of Arts recorded that there was a substantial increase in the number of working-class members in many institutes (ibid.: 180). As in Britain, institutes were providing education for the masses until they were superseded by other kinds of institutes, including establishing schooling for all and the establishment of working men's colleges during the 1880s and technical colleges at the beginning of the twentieth century. With this infrastructure in place, the Australian Mechanics' Institute Movement dominated the whole of the nineteenth century and, as Whitelock notes, their influence was from 12,000 miles away. However, by the beginning of the twentieth century, public education was established in most towns and the institutes, especially in the bush, became community centres and lending libraries. Many exist still, at the heart of their communities.

Canada

Ferry (2008a) makes a comparison between the Mechanics' Institute Movement in Britain and that in Canada, with the former relating to providing educational opportunities and the latter more community–related activities to support a broader need for the public good. He does state that the institutes found in the urban areas imitated those that were established in Britain.

The first mechanics' institutes established in Canada were to be found at Montreal (1828), York (1830) and Quebec (1831). Both had as their aim to support operatives in 'the instruction of Mechanics at a cheap rate in the principles of Arts which they practice, as well as in all other branches of useful knowledge' in support of the country's economic prosperity (Ferry 2008b: 443). Others were established at Toronto, Ottawa, Hamilton, Mitchell, Niagara, Paris and further afield. By 1892, there were 268 mechanics' institutes in Ontario (Wilson, Stamp *et al.* 1970: 231).

There is no doubt that the Canadian mechanics' institutes were based on those established in Britain, in support of skilled working-class men and after the 1850s women too, in providing adult education and libraries for their membership. However, it was only a few years before they 'proved to be adept to acclimatizing to colonial society' specific for a New World with emerging industrial but largely agricultural societies (Ferry 2008a: 463).

Ferry (2008b: 63) believes that the rural mechanics' institutes in Britain did not provide 'formalized education training and unsystematically formed mutual improvement societies'. Whereas in Canada he identified that rural institutes were more organised, offering subscription libraries, lectures and night classes, in Britain, as previously discussed, most small rural institutes were established with libraries at their centre. However, to be fair, rural institutes in Canada, as elsewhere, would be more isolated from towns than those in Britain and therefore were providing a more community–based foundation for their small population.

Like Britain, mechanics' institutes in Canada, particularly the rural ones, had seen a rise in female membership. However, unlike the British Movement, the institutes that could afford to do so added female reading rooms, such as at the Orillia Mechanics' Institutes (Ferry 2008b: 82), thus segregating men from women as was reflected in Canadian society generally in the nineteenth century.

Throughout the nineteenth century, the mechanics' institutes in Canada, and particularly those in rural settings, were open to all members of their villages and settlements, men, women, farmers, the self-employed and those in professional occupations. As Ferry states, 'Mechanics' Institutes managed to persist in their efforts towards mutual improvement and rational recreation until they were legalised completely out of existence by municipal directives in Quebec and by the *Free Library Act* of 1895 in Ontario'. In many ways this was similar to what happened in Britain, unless they became technical schools. By 1900, mechanics' institutes were in rapid decline. They had served their purpose, but with expanding industrialisation technical education required substantial support from the State and, as in Britain, free libraries, some founded by the American Carnegie Foundation, were being established and technical schools and colleges replaced the Institutes. (Wilson, Stamp *et al.* 1970). Unlike those in Britain, few seemed to have continued as early technical schools or colleges. To what extent was this partly due to the first Royal Commission on Industrial Training and Technical Education only being set up in 1910, compared to those in Britain following the Great Exhibition in 1851? It indicates that Canadian industrial development came later. With institutes now either public or subscription libraries, the Canadian government funded technical education through providing $3 million annually in support of state adult education (Goulson 1981: 339).

New Zealand

Between the 1840s and 1850s, with the establishment of settlers in New Zealand, there was an awareness of the need for education including support for adults, ideology brought from England by the New Zealand Company. Indeed the Company supported British values of education, and this included support for establishing mechanics' institutes. To establish institutes for self-improvement, with libraries, under physical hardships and challenges, was an amazing achievement. While the ideas for the Mechanics' Institute Movement had come from Britain, as in Australia the New Zealand institutes were more like community centres with libraries, rather than following the Birkbeck ideology of improving technical education (Hall 1970: 31).

Similarly, as with Australia, the long voyage of three months to New Zealand provided the opportunity for settlers to occupy their time with learning. However, Company policy insisted that few travellers were illiterate. The ships carrying new settlers brought boxes of books and had the ambition to establish a library in every settlement. This was true in the case of Nelson in 1841, and soon afterwards a mechanics' institute was established and the two were merged (ibid.). In 1846, the Richmond Mechanics' Institute was founded and held its meetings in a Wesleyan chapel. The timetable offered at the Institute included public readings on Saturday and Monday evenings, and elementary education on Wednesday and Friday evenings (ibid.: 31–2).

Hall (1970) believed that some institutes also had 'Athenaeum' in their title to give them 'greater dignity of intention' (ibid.: 33). Others were given the title 'Literary Institute'. Whatever title they had, their main aim was to offer a reading room and a library. Other institutes were established in New Zealand, including one at Wellington, Port Nicholas and Oamaru, which benefited from being near a gold mining region and had its own building by the 1860s. Lectures, not surprisingly, were often related to the mining industry. For example, in 1869, lectures were delivered on 'Watt and the Steam Engine' and 'Geology' (ibid.: 35). In 1869, the University of Otago was opened and professors often travelled to give lectures at Oamaru on their academic disciplines: chemistry, philosophy, classics, natural sciences. Timaru and Dunedin mechanics' institutes were reported in the 1879 Royal Commission as offering adult education (ibid.: 36).

With the passing of the Public Libraries Act in 1869, the Public Libraries Powers Act in 1875 and the Public Libraries Act of 1877, as well as new libraries many mechanics' institutes benefited from government subsidy. However, most became centres of their communities for entertainment and subscription libraries, as with such small populations there was not the opportunity to establish further education as there was in Britain. Thompson (1945: 23) thinks there were over 80 institutes by 1889. For the size of population in New Zealand at the time this was most impressive.

Gordon (2007) believed that Europe was way ahead of America by the 1880s with regard to technical education, reflected in the establishment of the City and Guilds London Institute for the Advancement of Technical Education in 1880 and the passing of the Technical Instructions Act in 1889 (ibid.: 25). While other countries also passed similar Acts, notably Australia, they did not have the same impact on further education as they did in Europe. It was only in 1910 that the State of Victoria in Australia, for example, passed an Education Act supporting free primary, secondary, technical and university education (Bessant 1974). However, Germany was far more advanced than any other country by the 1880s in support of technical education. For example, in Berlin in 1884 a technical high school was founded. The purpose of such schools and universities were to advance technical education so that Germany could become both economically and politically more advanced than Britain (ibid.: 25).

Conclusion

France, Germany, Italy, America, Canada, India, Australia and even some 'remote regions of the earth' including Honolulu, the principal port of the Sandwich

Islands (now known as the Hawaiian Islands), had established mechanics' institutes by 1850 (Hudson 1851: 214–21). In some countries the institutes were founded at the same time as those first established in Britain, circa 1824, while settlers in Australia took the ideas of the movement from Britain to Australia, with the first institute being established in Hobart, and then spreading across Victoria and beyond. Today, the mechanics' institutes are flourishing in parts of Australia, particularly Victoria, where they have become centres of the social life of rural settlements, or libraries in the larger towns such as Melbourne.

In all countries mentioned in this chapter, Birkbeckian influences have been clearly identified, if in some instances more British than others. That passion that useful knowledge should be available to all was taken up by Joseph Howe in Canada, Jonas Woodward in New Zealand, Josiah Holbrook in America and Henry Carmichael in Australia (Whitelock 1974: 87). While mechanics' institutes throughout the world had much in common, initially at least, with Glasgow and London, over time they naturally developed according to their own in-country needs and values for providing adult education.

Figure 10.3 As in Britain, various specialist institutes were established, such as railway institutes. The Victorian Railway Institute, situated near Melbourne, still survives and contributes to the local community. The building shown was built in 1928 and replaced an earlier railway building in 1917. Seymour was a major railway and army town and was an important regional centre. At its peak, about 600 people were employed in administration, railway line maintenance, the extensive workshops and catering. It was a railway stop for passengers travelling by train between Sydney and Melbourne. The Institute had an extensive technical library and lectures were held there for the various qualifications. General books were sourced from the very large general institute library in Melbourne. The workshops closed in the 1970s.

Source: Author's collection.

Note

1 The British referred to such fairs as exhibitions and they were basically the same in their aims, raising money for the local institute, exhibiting goods, art and machines.

References

Baragwanath, P. (2000) *If The Walls Could Speak: A Social History of The Mechanics' Institutes of Victoria* (Mechanics' Institutes of Victoria).

Bessant, B. (1974) 'The Australian Labour Movement and education prior to 1914', *Journal of the History of Education Society* Vol. 3, No. 2, 40–56.

Bode, C. (1956) *The American Lyceum, Town Meeting of the Mind* (New York: Oxford University Press).

Clancy, F. (2000) *The Libraries of the Mechanics' Institutes of Victoria*, report prepared for the Department of Infrastructure (Victoria: Department Publication).

Doug, J. (1991) 'William Maclure and the New Harmony Working Men's Institute', *Journal of Libraries and Culture* Vol. 26, No. 2, 402–414.

Dunn, M. (2012) 'Technical and Working Men's College', *Dictionary of Sydney*. Available online at www.dictionaryofsydney.org/entry/technical_and_working_mens_college (accessed February 2015).

Ferry, D. (2008a) 'Culture, authority, and the emergence of a liberal social order in the Central Canadian Mechanics' Institute Movement, 1828–1860', Christie, N. (ed.) *Transatlantic Subjects, Ideas, Institutions and Social Experience in Post–Revolutionary British North America* (Montreal & Kingston: McGill-Queen's University Press): 439–474.

Ferry, D. (2008b) *Uniting in Measures of Common Good: The Construction of Liberal Identities in Central Canada, 1830–1900* (Montreal & Kingston: McGill-Queen's University Press).

Goldsmith, J. S. (1955) *One Hundred Years of Mechanics' Institute of San Francisco: 1855–1955* (San Francisco: Greenwood Press).

Goulson, C. F. (1981) *A Source Book of Royal Commissions and Other Major Government Inquiries in Canadian Education 1787–1978* (Toronto: University of Toronto).

Gordon, D. R. (2007) *Rochester Institute of Technology: Industrial Development and Educational Innovation in an American City 1829–2006* (New York: Rochester Institute of Technology Press).

Hall, D. O. W. (1970) *New Zealand Adult Education* (London: Michael Joseph).

Hudson, J. W. (1851) *The History of Adult Education in which is comprised a Full and Complete History of the Mechanics' and Literary Institutions* (Reprint. London: Woburn 1969).

Kett, J. F. (1994) *The Pursuit of Knowledge under Difficulties: From Self-Improvement to Adult Education in America, 1750–1990* (Stanford, California: Stanford University Press).

Nadel, G. (1957) *Australia's Colonial Culture: Ideas, Men and Institutions in Mid-Nineteenth Century Eastern Australia* (Melbourne: F. W. Cheshire Press).

Ray, A. G. (2005) *The Lyceum and Public Culture in the Nineteenth-Century United States* (East Lansing, Michigan: Michigan State University Press).

Rice, S. P. (April 2000) 'The Mechanics' Institute of the City of New York and the conception of class authority in early industrial America, 1830–1860', *New York History* (New York: New York State Historical Association): 261–298.

Ryan, S. L. (1974) *The Development of State Libraries and their Effect on the Public Library Movement in Australia 1809–1964* (Queensland: Library Board of Queensland).

Sinclair, B. (1974) *Philadelphia's Philosopher Mechanics: A History of the Franklin Institute 1824–1865* (Baltimore: The John Hopkins University Press).

South Bourke Standard (25 October 1861) (author unknown).

Stam, D. H. (ed.) (2001) *International Dictionary of Library Histories* Vol. 2 (Chicago: Fitzroy Dearborn).

Stevens, E. W. (Winter 1990) 'Technology, literacy and early industrial expansion in the United States', *History of Education Quarterly* Vol. 30, No. 4, Special Issue on the History of Literacy, 523–544.

Talbot, M. (1992) *A Chance to Read: A History of the Institutes Movement in South Australia* (Adelaide: Libraries Board of South Australia).

Thompson, A. B. (1945) *Adult Education in New Zealand* (New Zealand Council for Educational Research: Dunedin).

Tylcote, M. (1957) *The Mechanics' Institutes of Lancashire and Yorkshire Before 1851* (Manchester: Manchester University Press).

Whitelock, D. (1974) *The Great Tradition: A History of Adult Education in Australia* (Queensland: University of Queensland Press).

Wiegand, W., and Davis, D. (1994) *Encyclopaedia of Library History* (New York: Garland Publishing).

Wilentz, S. (1984) *Chants Democratic: New York City and the Rise of the American Working Class, 1788–1850* (Oxford: Oxford University Press).

Wilson, J. D., Stamp, R. M., and Audet, L. (1970) *Canadian Education: A History* (Ontario: Prentice–Hall of Canada).

11 Conclusion and legacy of the Mechanics' Institute Movement

Introduction

The research has identified four key areas that reveal that mechanics' institutes provided a major contribution to working-class adult education between 1824 and circa 1900. The discussion has concentrated on class and female membership, technical education as a result of the Great Exhibition and foreign competition, and curriculum developments. Other indicators, including membership patterns and accommodation developments, have also been highlighted, to further point towards their success. The Mechanics' Institute Movement was not declining by 1850, as Hudson feared. By taking a longer view than most historians have, this book argues that the Movement continued to grow and be successful after 1851. Although scientific lectures and classes were offered throughout the period, the introduction of elementary education at mechanics' institutes, for children as well as adults, and relevant scientific and technological subjects relating to industry, encouraged and developed a much wider membership. The result was that the mechanics' institutes often had to move into larger rented accommodation or build their own, which many did. While the founders were often from the wealthy classes, the majority of members were working men and women who often found themselves mixing with the professional classes, in the larger town institutes at least. Individuals, such as George Birkbeck, made important contributions to the success of the Mechanics' Institute Movement. Radicals, particularly the Unitarians, supported the education of both men and women of the working population, supporting a relevant curriculum that was required to support semi-rural and urban industrialisation. Indeed, presidents of mechanics' institutes were often Unitarians themselves. The Movement was given national credibility through public recognition offered by the Society of Arts, who organised national examinations in the local institutes. This fitted well with those who used mechanics' institutes. Qualifications validated the efforts of those who sought to 'better themselves'. The number of patents licensed in mechanics' institute towns is evidence that the institutes were offering relevant curricula to support technological developments and progress. Finally, specific research in relation to the Yorkshire Union of Mechanics' Institutes has put the national Mechanics' Institute Movement into a regional and local context. The research has supported the hypothesis that mechanics' institutes continued to be successful, and their numbers grew after 1850 up to and beyond the passing of the technical instruction acts, which were to provide a firm base on which state-funded adult education would be established. The order of the chapters in this volume has provided historical chronology with regard to the growth and development of the Mechanics' Institute Movement and national events that had an impact on it.

As has been highlighted in Chapter 1, the history of education not only provides insight into the educational past but also helps to explain how present-day developments have been established as well as having a deeper understanding of the function that education fulfils today. It was also emphasised that the history of education has been somewhat neglected by both historians and educationalists but that a study of this kind emphasises the importance of educational history to other subjects, or 'neighbours', most obviously history but also sociology, politics, culture, architecture and geography. All these underpin educational developments and influences during the period of study with regard to the beginnings of adult working-class education and, in particular, the contribution made by the Mechanics' Institute Movement, which also included similar institutions such as literary and scientific societies. The Movement went on to become the foundation on which the further education sector was introduced in the early twentieth century. Later, the Movement also influenced higher education through the establishment of colleges of technology and polytechnics, many of which could trace their origins to their local institute, and which went on to become universities.

The research has identified several areas that reveal that mechanics' institutes and similar provided a major contribution to working-class adult education, something previously neglected. Discussion has concentrated on class and female membership, technical education and curriculum developments. Other indicators, including membership patterns and accommodation developments have also been discussed and analysed to further point towards the success of the Movement. The work has emphasised that mechanics' institutes were not declining by 1850, as J. W. Hudson feared, reinforced by historians, and by taking a longer view, up to the end of the nineteenth century, than educationalists and historians have done, I argue that the Mechanics' Institute Movement continued to grow after 1851 and contributed to working-class adult education. After a rather uncertain start, during which several institutes closed or declined in importance due to declining membership, the Movement responded by offering more relevant subjects in supporting a working-class educated workforce as a response to rapid industrialisation and foreign competition.

While the founders were often from the wealthy classes, over time and with the introduction of elementary education and relevant curricula, some of the committee members were themselves working class. In rural and semi-rural areas where towns were expanding, the institutes were able to offer relevant subjects to their mainly working-class population.

The Mechanics' Institute Movement was given national credibility through public recognition offered by the Science and Art Department, South Kensington, the Society of Arts and later the City and Guilds London Institute, who offered national examinations. Qualifications validated the efforts of those who sought to better themselves and gain recognition for their studies.

Specific research in relation to the Yorkshire Union, using the surviving annual reports, has put the national Mechanics' Institute Movement into both a regional and local context. The research has supported the hypothesis that mechanics' institutes continued to be successful and their numbers grew after 1850 up to and beyond the passing of the Technical Instruction Acts, which were to provide a firm base on which state-funded adult education would be established. The order of the chapters provides historical chronology and analysis with regard to the growth and development of the Mechanics' Institute Movement and national events that had an impact on it.

In many ways, mechanics' institutes had evolved very much as a result of trial and error. The curricula initially offered were ambitious, as they offered advanced-level scientific subjects and lectures. Many institutes closed, such as those at Skipton and Huddersfield, but later reopened after rebranding themselves for a wider membership, while others, notably Manchester, saw a rapid decline in membership. Committees realised the importance of establishing elementary education for all males and females, identifying that such knowledge was the foundation for education, in particular in relation to industrialisation, on which more advanced technical education could be built. Science, particularly chemistry in relation to the textile industry, steam, art subjects, such as industrial design in relation to machine construction and, in rural areas, agricultural subjects were introduced to support advanced-level learning and supported relevant employment in textiles, engineering coal mining, railways and agriculture, to name but just a few. University extension courses and lectures were offered at many institutes, relating to these subjects.

McCulloch (1997: 69) observed that:

> education policies have often appeared to be lacking in historical perspective. The events and problems of the past tend to be treated as an irrelevant distraction to the problems at hand. Unmistakably, however, history continues to impinge on even the most historically unaware of education policies.

The passing of Forster's Education Act of 1870 provided the beginnings of state education for children until the age of 13 years, and this had been partly due to the government's fear that Britain would lose its industrial position without a future educated workforce. Such concerns continue to resonate today. The institutes were ideally positioned to support the Act, which gradually not only made elementary education compulsory but also supported opportunities for those who could to gain higher-level qualifications previously unobtainable for the masses. It was rather fitting that W. E. Forster, Member of Parliament for Bradford in Yorkshire and the Minister responsible for drawing up the 1870 Act, was a supporter of the Movement nationally and particularly across Yorkshire. He donated funds to several individual institutes and was on several committees. As O'Farrell (2004: 164) has observed, 'had the founders of mechanics' institutes succeeded in persuading governments to take up the cause of mechanics' institutes earlier, the history of nineteenth-century technical education and indeed Britain's industrial progress might have been different'. The Royal Commissions on technical education and the resulting Acts were influenced by the Movement that resulted in more state-funded technical education by the end of the nineteenth century.

It was the Great Exhibition of 1851 that first alerted the government to the need for an educated workforce and which, through the Mechanics' Institute Movement, had made this possible. The Science and Art Department and Society of Arts examinations provided the opportunity for members to qualify in a range of subjects associated with technical education. Royal Commissions resulted in the passing of the Technical Instruction Act of 1889, which was the first Act to support adult education. The Local Tax Act of 1890 raised much-needed revenue to support adult education. Under the 1870 Act, local authorities managed the process so towns not able to provide enough places for children were highlighted as areas for School Boards to be set up. Local education authorities replaced the 300 school boards under the Education Act of 1902, which also became responsible for secondary and adult education, the latter becoming part of the twentieth-century further education sector.

Kelly (1973: 18) makes the point that the Mechanics' Institute Movement did lay the foundations on which technical education and public libraries were established. This is further underlined by Venables (1956: 14), who states that technical education began with the Mechanics' Institute Movement from 1823.

Nevertheless, 'had the founders of mechanics' institutes succeeded in persuading governments to take up the cause of mechanics' institutes earlier, the history of nineteenth century technical education and indeed Britain's industrial progress might have been different' (O'Farrell 2004: 164).

With the passing of the Technical Instruction Acts in 1889 and 1890, the government finally took over the funding of technical education. It was the Local Taxation (Customs and Excise Tax) Act that enabled local authorities to levy certain taxes, principally on alcohol. This was often referred to as 'whisky money', and was a key source of funding for local technical education. The phrase 'to distil wisdom out of whisky, genius out of gin and capacity for business out of beer' was referred to by Members of Parliament in the House of Commons in informing the town, county and county borough councils of the need for them to promote technical education (Venables 1956: 22). By the turn of the century, whisky money contributed more to technical education budgets than grants from the Education department or local authority rates. Finally, at the end of the nineteenth century, the *Report of the Royal Commission on Secondary Education* (The Bryce Report), published in 1895, recommended a Minister for Education taking over the Education Department, the Science and Art Department and the Charity Commission. The Minister would have responsibility for universities, schools and local education authorities. It was under these powers, supported by the Education Acts of 1902, 1918 and 1944, that the country saw the decline and eventual extinction of mechanics' institutes, although their buildings were often converted into museums and free public libraries. They were replaced by schools of art and technical colleges for post–school age students (Maclure 1972: 140).

Although scientific lectures and classes were offered throughout the period, the introduction of elementary education particularly from the 1840s, and later for children as well as adults, as well as relevant scientific and technological subjects relating to industry, encouraged and supported a much wider membership. The result was that many mechanics' institutes, which had been established in small rented buildings, sometimes a single room, had to move into larger rented space, with many moving into their own purpose-built accommodation. Committees were able to provide larger classrooms and laboratories and gave institutes the opportunity to provide larger libraries and reading rooms and to introduce penny savings banks. Some institutes had chemistry laboratories that were probably more advanced than those in many universities. Design reflected the Victorian Gothic style, which contributed to the civic pride of the town, while smaller institute buildings were often influenced by Nonconformist chapels and board schools.

Focusing on practical technical and academic learning in support of the local economy was the main aim of all mechanics' institutes, something that the further education sector has subsequently continued to respond to. The vocational and training qualifications offered through the Royal Society of Arts (it gained 'Royal' status in 1900) became Oxford Cambridge RSA (OCR), offering commercial examinations as well as General Certificates in Secondary Education (GCSEs) sat by 16-year-olds, and General Certificates in Education (GCE) 'A' Levels for entry into university. The City of Guilds London Institute still offers qualifications in technical, vocational diplomas and apprenticeships. The university technical colleges (UTCs), introduced between 2010 and 2016, are in many ways similar to those mechanics' institutes that offered advanced work-based courses, but relevant for the twenty-first century economy. They offer GCSEs, 'A' Levels, City and Guilds foundation and full degrees through a sponsored university. Like mechanics' institutes, they are taking some time to research their potential, with few meeting their annual student number intakes.

The Mechanics' Institute Movement responded to the needs of the working class though adopting a relevant curriculum, introducing qualifications, providing library access and responding to national challenges over a period of 76 years or so and preparing the way forward in offering post–school education for all. As Hyland and Merrill (2003: 6–7) note, mechanics' institutes resulted in the beginning of 'a development which was to lead to the establishment of technical colleges' at the end of the nineteenth century and which provided the foundation on which the further education sector was established.

The Mechanics' Institute Movement therefore provided a firm foundation on which the development in early twentieth-century technical and further education was established in towns. These included Birmingham, Blackburn, Bolton, Bristol, Burnley, Cardiff, Derby, Dewsbury, Halifax, Huddersfield, Keighley, Leeds, Liverpool, Manchester, Portsmouth, Preston, Rochdale, St Helens, Salford, Southampton, Swindon, West Bromwich, West Ham, Westminster, Wolverhampton and Wigan, to name but a few.

At Leeds, the School of Art was opened in 1846 and, having been funded by the Institute, was attached to the original Institute. In the case of Keighley, following a serious fire during the 1960s, rather than demolishing the surviving part of the Institute (see Figure 7.1 with the Weaving School Wing on the left), one of the then-new college buildings was connected to the remaining building. While most of the frontage was destroyed, including the clock tower, the back section and the Weaving School Wing on the left survived (see Figure 11.1). The iconic building, physically merging the old with the new, was closed, and a new twenty-first century college was built in another part of the town.

These towns had previously had active mechanics' institutes or similar prior to government funding being available for technical education. There was further

Figure 11.1 Keighley Further Education College, located on the high street. The photograph was taken from the same position as the illustration in Figure 7.1. Note the gable end of the original building, the Weaving School, which was partly funded by Worshipful Company of Clothmakers.

expansion in the number of technical colleges from 1902 onwards, many having originally been mechanics' institutes (Venables 1956: 32). Their foundations were based on working-class adult education and community learning, providing the opportunity for young people on leaving school to progress into further education.

Mechanics' institutes specialised in supporting their local industry and commerce and these traditions were continued and developed by their successive technical collages. Wigan Mechanics' Institute, for example, became the Wigan and District Mining and Technical College and Barnsley Mechanics' Institute became Barnsley Mining College. Leeds College of Art can trace its developments back to Leeds Mechanics' Institute, one of the most successful of its kind in the country and headquarters of the Yorkshire Union. Huddersfield Mechanics' Institute, which had a national reputation for offering chemistry at various levels relating to the textile-dyeing industry, became a college of technology, polytechnic and, in 1994, a university. The former Mechanics' Institute and Technical School building is now part of the University of Huddersfield's campus. Several others, such as Manchester University, can trace their origins back to the Mechanics' Institute Movement, as can the former Anderson Institution, which went on to become the Glasgow Royal Technical College and which is now Strathclyde University (O'Farrell 2004: 19). The London Mechanics' Institution became Birkbeck College, part of the University of London in 1907 (ibid.). Salford Mechanics' Institute, founded in 1858, and nearby Pendleton Mechanics' Institute, established in 1850, were merged and became the Technical Institute, continuing to offer subjects first introduced by the individual institutes: engineering, electrical engineering, physics, chemistry, dyeing, spinning and weaving. In 1921, the Institute became the Technical College, Salford, and in 1967 it received its Royal Charter and became the University of Salford.

As highlighted in Chapter 3, Robert Elliot (1861) observed that 'the banquet was prepared for guests who did not come', highlighting the view of several educationalists and historians that although mechanics' institutes were founded on behalf of the working classes, they were reluctant to attend. The research for this book has gone some way to argue that this was not the case and that guests did arrive at the banquet, albeit in small numbers to begin with. The Mechanics Institute Movement made a substantial contribution to both elementary and technical education as well as providing a firm foundation on which further and higher education has subsequently been developed.

References

Elliot, R. (1861) 'On the working men's reading rooms, as established in 1848 at Carlisle', *Transactions of the National Association for the Promotion of Social Science*: 110–117.

Hyland, T., and Merrill, B. (2003) *The Changing Face of Further Education; Lifelong Learning, Inclusion and Community Values in Further Education* (London: RoutledgeFalmer).

Kelly, T. (1973) *A History of Public Libraries in Great Britain 1845–1965* (London: The Library Association).

Maclure, J. S. (1972) *Educational Documents England and Wales 1816–1968* (London: Methuen).

McCulloch, G. (1997) 'Privatising the past? History and education policy in the 1990s', *British Journal of Educational Studies* Vol. 45, No. 1, 69–82.

O'Farrell, P. N. (2004) *Heriot-Watt University: An Illustrated History* (Edinburgh: Pearson Education Ltd).

Venables, P. F. R. (1956) *Technical Education: Its Aims, Organisation and Future Development* (London: Bell and Son).

Index